Catalytic Activation
of Carbon Dioxide

ACS SYMPOSIUM SERIES **363**

Catalytic Activation of Carbon Dioxide

William M. Ayers, EDITOR
Electron Transfer Technologies

Developed from a symposium sponsored
by the Division of Colloid and Surface Chemistry
at the 191st Meeting
of the American Chemical Society,
New York, New York,
April 13–18, 1986

American Chemical Society, Washington, DC 1988

Library of Congress Cataloging-in-Publication Data

Catalytic activation of carbon dioxide/William M. Ayers, editor.

p.   cm.

"Developed from a symposium sponsored by the Division of Colloid and Surface Chemistry at the 191st meeting of the American Chemical Society, New York, New York, April 13-18, 1986."

ISBN 0-8412-1447-6

1. Carbon dioxide—Congresses. 2. Catalysis—Congresses.

I. Ayers, William, 1950-   . II. American Chemical Society. Division of Colloid and Surface. III. American Chemical Society. Meeting (191st: 1986: New York, N.Y.)

QD181.C1C38   1988
665.7'7—dc19   87-30832
                                                 CIP

ACS Symposium Series

M. Joan Comstock, *Series Editor*

1988 ACS Books Advisory Board

Foreword

The ACS SYMPOSIUM SERIES was founded in 1974 to provide a medium for publishing symposia quickly in book form. The format of the Series parallels that of the continuing ADVANCES IN CHEMISTRY SERIES except that, in order to save time, the papers are not typeset but are reproduced as they are submitted by the authors in camera-ready form. Papers are reviewed under the supervision of the Editors with the assistance of the Series Advisory Board and are selected to maintain the integrity of the symposia; however, verbatim reproductions of previously published papers are not accepted. Both reviews and reports of research are acceptable, because symposia may embrace both types of presentation.

Contents

Preface..ix

1. **Sources and Economics of Carbon Dioxide**...........................1
   Sol J. Barer and Kenneth M. Stern

2. **Carbon Dioxide Equilibria**...8
   James N. Butler

3. **Coordination of Carbon Dioxide to Nickel: An Alternative Theoretical Model**...16
   R. P. Messmer and H.-J. Freund

4. **Metal-Induced Transformations of Carbon Dioxide**....................26
   Donald J. Darensbourg, Christopher G. Bauch, and Cesar Ovalles

5. **Use of Stoichiometric Reactions in the Design of Redox Catalyst for Carbon Dioxide Reduction**...................................42
   Daniel L. DuBois and Alex Miedaner

6. **Electrocatalytic Carbon Dioxide Reduction**..........................52
   B. Patrick Sullivan, Mitchell R. M. Bruce, Terrence R. O'Toole, C. Mark Bolinger, Elise Megehee, Holden Thorp, and Thomas J. Meyer

7. **Enzymatic Activation of Carbon Dioxide**.............................91
   Leland C. Allen

8. **Adsorptive and Catalytic Properties of Carbon Monoxide and Carbon Dioxide Over Supported Metal Oxides**............................102
   D. G. Rethwisch and J. A. Dumesic

9. **Adsorption and Reaction of Carbon Dioxide on Zirconium Dioxide**.......123
   Ronald G. Silver, Nancy B. Jackson, and John G. Ekerdt

10. **Effect of Potassium on the Hydrogenation of Carbon Monoxide and Carbon Dioxide Over Supported Rh Catalysts**.......................133
    S. D. Worley and C. H. Dai

11. **Carbon Dioxide Reduction with an Electric Field Assisted Hydrogen Insertion Reaction**...147
    W. M. Ayers and M. Farley

12. **Electrochemical Reduction of Aqueous Carbon Dioxide at Electroplated Ru Electrodes: Investigations Toward the Mechanism of Methane Formation**..155
    Karl W. Frese, Jr., and David P. Summers

13. **Electrochemical Studies of Carbon Dioxide and Sodium Formate in Aqueous Solutions**...171
    M. H. Miles and A. N. Fletcher

14.   Electrochemical Activation of Carbon Dioxide 179
      K. Chandrasekaran and J. O'M. Bockris

INDEXES

Author Index **206**

Affiliation Index **206**

Subject Index **206**

Preface

INCREASING USE OF FOSSIL FUELS (petroleum, coal, and gas) will continue to load the atmosphere with carbon dioxide beyond the apparent capacity of the plant and oceanic sinks to absorb the gas. There has been an estimated 15% increase in atmospheric carbon dioxide since the turn of the century (1). An active response to this worldwide problem is to capture and chemically convert the carbon dioxide at its source of production. Ironically, these conversion processes must be powered by nonfossil fuel sources (solar or nuclear) to achieve a net reduction in atmospheric carbon dioxide.

The expected return of oil price increases over the next decade will again spur investigation into conversion processes for alternative fuel sources such as coal. Coal gasification will produce substantial amounts of carbon dioxide as a by-product. If this carbon dioxide could be converted economically to methanol or methane, established zeolite catalytic processes could convert these intermediates to gasoline. Furthermore, concern about providing a substitute for natural gas to the established gas pipelines has led the Gas Research Institute to sponsor investigation of carbon dioxide conversion selectively to methane (2).

This volume is based on a symposium that is part of a continuing series in the Surface Science of Catalysis, sponsored by the Division of Colloid and Surface Chemistry of the American Chemical Society. The symposium was motivated by an interest in C_1 chemistry and the desire to convert an abundant material, carbon dioxide, into useful products.

Investigation of carbon dioxide catalytic activation is explored by a variety of subdisciplines (homogeneous catalysis, heterogeneous catalysis, electrocatalysis/photoelectrocatalysis), often with little cross-citation of work. This situation created a need to bring together the leading researchers to provide an overview of methods and accomplishments to date.

The papers range from general issues (sources, economics, and physical properties) to theoretical treatments of carbon dioxide bonding, followed by the various catalytic approaches (homogeneous, enzymatic, heterogeneous, electrocatalytic, hybrid, photoelectrocatalytic). The volume does not cover noncatalyzed photochemical reactions or energetic particle (electron beam and fission fragment) activation of carbon dioxide. These methods and an overview of earlier approaches for converting carbon dioxide can be found in a series of reports prepared at Brookhaven National Laboratory (3).

Realistically, both carbon dioxide and carbon monoxide need to be examined as feedstocks for future applications. These two materials are linked together through the water gas shift reaction:

$$CO + H_2O = CO_2 + H_2$$

The recent advances in catalysis of the shift reaction have been reviewed by Ford (4), Laine (5), and in a companion volume in the Symposium Series, *Catalytic Activation of Carbon Monoxide* (6).

Prior to this symposium, reviews of homogeneous (7, 8) and electro-catalytic (9) activation of carbon dioxide suggested its emergence as an alternative to carbon monoxide, primarily on the basis of its low cost, abundance, and lower toxicity. The challenge with carbon dioxide reactions (in which hydrogen species act as oxygen acceptors) is to overcome the kinetic barriers to reaction.

I wish to thank the symposium participants, Robin Giroux of ACS Books, and The Carbon Dioxide Research Division, Office of Basic Energy Sciences, Office of Energy Research of the U.S. Department of Energy for financial support through Grant No. DE-FG05-87ER13760.

Literature Cited

1. *Carbon Dioxide and Climate*; U.S. Department of Energy. Office of Basic Energy Sciences, 1984; DOE/ER-0202.
2. Cook, R. L.; MacDuff, R. C.; Sammells, A. F. *J. Electrochem. Soc.* **1987,** *134,* 2375.
3. Steinberg, M. *An Analysis of Concepts of Controlling Atmospheric Carbon Dioxide*; 1984; BNL-33960.
4. Ford, P. C. *Acc. Chem. Res.* **1981,** *14,* 31-37.
5. Laine, R. M.; Wilson, R. B. *Aspects Homogeneous Catal.* **1984,** *5,* 217-40.
6. *Catalytic Activation of Carbon Monoxide*; Ford, P. C., Ed.; ACS Symposium Series No. 152; American Chemical Society: Washington, DC, 1981.
7. Darensbourg, D. J.; Ovalles, C. *CHEMTECH* April **1985,** 636-640.
8. Denise, B.; Sneeden, R. P. A. *CHEMTECH* February **1982,** 118-112.
9. Ulman, M.; Aurian-Blajeni, B.; Halmann, M. *CHEMTECH* April **1984,** 235-239.

WILLIAM M. AYERS
Electron Transfer Technologies
Princeton, NJ 08542

September 23, 1987

Chapter 1

Sources and Economics of Carbon Dioxide

Sol J. Barer[1] and Kenneth M. Stern

Chem Systems, Inc., Tarrytown, NY 10591

In order for carbon dioxide to be
considered as a significant feedstock for
the chemical industry it must be both
readily avialable and economically
attractive. Sources of carbon dioxide can
be catagorized in terms of tne
concentration of this material - high or
low. Examples of the former include
natural reservoirs, natural gas processing
plants and facilities engaged in the
production of ethylene oxide, ammonia or
hydrogen. The largest potential supply of
carbon dioxide, however, is from the
dilute sources which comprise various
fossil fueled power plants (including
coal, oil-and gas-fired facilities).
This paper will discuss some of the issues
associated with the various sources as
well as provide a perspective on their
economics.

Carbon dioxide is an industrial gas with two distinct
identities. Its traditional role relates to recovery from
ammonia production and use in urea manufacture. Smaller
amounts are recovered from other sources (e.g., ethanol
plants), and the merchant CO_2 market also includes
consumption in end uses such as refrigeration and beverage
carbonation. An exciting potential, of course, is for use
as primary petrochemical feedstock.

---

[1]Current address: Celgene, 7 Powder Horn Drive, Warren, NJ 07060

Conventional uses for CO_2 are:

1. Urea (approximately 5200 thousand metric tons/year) -
    this is basically a captive market.
2. Refrigeration (approximately 1500 thousand metric
    tons/year).
3. Beverages (approximately 750,000 metric tons/year).
4. Enhanced oil recovery (approximately 6-7 million
    metric tons/year).

As detailed below the basic sources of CO_2 are natural
gas processing plants, ethanol plants, ammonia plants,
hydrogen plants and ethylene oxide plants. In addition,
there are a variety of dilute sources, the largest
category of which is power plants. As shown below, of the
man-made sources, the largest is ammonia production
approximately 1100 MCF (million standard cubic feet).
This is to be compared with approximately 80-90,000 MCF
from power plants. Natural sources total 1500-2000 MCF.
the economics for the use of carbon dioxide arising form
these high concentration sources is dependent on clean-up
and compression cost.

Natural Reservoirs

The major reservoirs of natural CO_2 occur around the
Permian Basin area. Sheep Mountain, in southeastern
Colorado is estimated to contain one trillion cubic feet
(TCF) of CO_2 of 97% purity. Productive capacity is
approximately 300 MCF per day. Other important sources
include Brano Dome in New Mexico with reserves of 5 TCF
and with total productive capacity of approximately 350
MCF per day. The McElmo Dome has reserves of greater than
8 TCF (97% purity) of the same productive capacity as the
Brano Dome Unit. The potential for the McElmo Dome is
believed to be approximately 1 billion cubic feet per day.
In addition, other units include Jackson Dome, Mississippi
(1 TCF proven) and the LaBrage area of southwestern
Wyoming which is believed to have reserves in excess of 20
TCF. These data, based on the 1984 National Petroleum
Council study of enhanced oil recovery, indicate that the
aggregate supply is approximately 2 billion cubic feet per
day.

Natural Gas Processing Plants

There are many deposits of natural gas which also
contain appreciable quantities of CO_2. The CO_2 can be
separated by any one of a number of acid gas removal
processes. The approximate CO_2 emission from the natural
gas processing plants is given in the table below:

Natural Gas Processing Plants Emitting CO_2

State	Approximate CO_2 Produced (MCF/Day)
Louisiana	6
New Mexico	12
Texas	210
W. Virginia	20
Wyoming	6
Total	254

Source: "Feasibility and Economics of By-Product CO_2 for Enhanced Oil Recovery", DOE contract DE-AT21-78, p.181-182 MC08333.

As is the case for natural CO_2 reservoirs, CO_2 recovered from natural gas processing plants is generally of sufficient quantity to be pipelined to oil fields for use in miscible flood cooperation. Carbon dioxide produced from this source is generally considered to be distinct from traditional merchant market sources.

Ethanol Plants

The sources of ethanol in the United States have been undergoing significant change in recent years. Up until the past few years, synthetic ethanol dominated the industrial ethanol market. Domestic fermentation ethanol capacity for industrial/fuel applications was negligible compared to the capacity for synthetic ethanol.

Over the past five years, the capacity of plants making fermentation ethanol primarily for fuel use has rapidly increased. Fermentation ethanol capacity in the United States is estimated to be approximately 850 million gallons per year.

In the fermentation process, CO_2 and ethanol are produced in roughly equal weights. This translates into a CO_2 production rate of about 60 SCF per gallon of ethanol. Based on the capacity figure cited above, total CO_2 supply from this source is estimated to be 130 MCF/D. Recovered CO_2 from this source would generally find application in the conventional CO_2 merchant market.

Ammonia Plants

Ammonia is produced by the catalytic reaction of nitrogen and hydrogen at high temperature and pressure. The overall process includes the major steps of desulfurization, reduction of feed with steam and reaction of reformed gas with air over catalyst. Subsequently, CO is reformed with additional hydrogen (via steam) and CO_2 is removed.

As a result of the shift reaction, significant quantities of CO_2 are produced during the process, which can be coupled with urea plants to use the CO_2 captively. U.S. ammonia plants have nearly 18 million tons of capacity corresponding to a CO_2 production rate of approximately 1 billion cubic feet per day. This (when subtracted from CO_2 use for urea) translate into an availability of approximately 800 MCF per day. Only about 3.5 million metric tons actually reach the market.

Hydrogen Plants

Hydrogen is produced in large quantities in many refineries for use in operations such as hydrotreating, isomerization and hydrocracking. It is most frequently obtained from a synthesis gas produced by steam reforming of natural gas.

In the production of hydrogen, pressurized natural gas is desulfurized over active carbon or hot zinc oxide, mixed with steam to give the required steam-to-feedstock carbon molar ratio (typically three to one) and then steam-reformed.

To generate a high purity hydrogen product, the CO_2 stream, after moisture removal, is either vented or sold. Approximately 25 SCF of CO_2 are produced for every 100 SCF of hydrogen. At a total U.S. capacity of approximately 2 billion cubic feet per day of hydrogen and approximately 550 MCF per day of CO_2 can be produced.

Ethylene Oxide

The conventional route to ethylene oxide entails the direct vapor phase oxidation of ethylene. The reaction proceeds at 200-300°C and 10-30 atmospheres to produce ethylene oxide in 65-80 mole percent selectivity. The success of this technology is attributable to the development of fairly selective silver oxide catalysts which limit combustion of ethylene to CO, CO_2 and water. The CO_2 is present in the purge gas. Oxygen-based ethylene oxide plants produce approximately 60 MCF of CO_2 per day.

Dilute Sources

Dilute sources of CO_2 consist of sources in which the CO_2 would otherwise be discharged to atmosphere in a flue gas stream. Relatively high expenses would be incurred for recovery of the CO_2.

To date, only one project has attempted to utilize the exhaust stream from an electric power plant for CO_2 recovery. This project involved recovery of CO_2 from the flue gas of a gas-fired utility boiler and delivered approximately 100 tons CO_2/day.

Coal-fired power plants have CO_2 exhaust content ranging from approximately 23,000 to 43,000 SCF/ton of coal. Oil-and Gas-fired Power Plants, produce approximately 10 billion cubic feet per day.

The estimated total production of CO_2 from power plants is quite large (approximately 90 BCF per day). However, there are a number of issues with respect to the use of such CO_2 including the presence of sulfur compounds in the gas which could seriously effect catalysts.

Summary of CO_2 Supply

Carbon dioxide is derived from natural and man-made sources, with the latter make up of both high concentration and low concentration sources. Of the total CO_2 generated, only a small portion is recovered for subsequent sale, whereas natural CO_2 is generally recovered for large volume applications such as enhanced oil recovery. However, for sheer magnitude, the CO_2 produced in power plants is many times larger than the other sources. Unfortunately, this huge volume of CO_2 is in dilute form necessitating concentration. From the perspective of the chemical industry, as opposed to those interested in the massive use of CO_2 for enhanced oil recovery, the quantities required would appear to be sufficient from the high concentration sources.

Recovery Technology

Carbon dioxide can be recovered in a number of ways. The choice of system depends on various factors, including:

- Feed gas CO_2 concentration and pressure
- Percent recovery required
- Presence of contaminants that may foul the equipment or solvents

Conventional carbon dioxide recovery systems fall into the following categories:

- Chemical solvent systems that chemically react with selected gases, regenerate the solvent by reversing the reaction with heat and/or pressure let down wherein absorbed acid gases are released. The solvents are typically alkanolamines or hot potassium carbonate.

- Physical solvent systems in which gas is absorbed by physical means in a solvent and released by heating and/or pressure let-down.

•   Membranes offering the unique property of
    selective permeability. These are permeable
    films that permit some gases to transfer through
    them more rapidly than others, thereby permitting
    components to be separated. They operate on a
    continuous flow basis.

•   Cryogenic separation in which very low
    temperatures are used to chill the feed streams
    and allow distillation of the various components.

Of these alternatives, the solvent absorptions
processes are by far the most frequently utilized for CO_2
recovery.

Economics of CO_2 Recovery

Estimation of CO_2 recovery economics is a complex
undertaking, primarily because of (a) the wide variations
in composition among the many potential feed streams, and
(b) the numerous available recovery techniques and
configurations thereof.

An important source of CO_2 for recovery is natural CO_2
deposits (e.g., in association with natural gas). Such
streams would exhibit a wide range of compositions; one
representative stream might be characterized as 50 percent
methane and 50 percent CO_2.

The cost of CO_2 recovery from this high level CO_2
stream via a triethanolamine (TEA) solvent process is
approximately $.50/MCF with the major single portion
(approximately 45%) being the utility component. For a
membrane system the cash cost is lower (approximately
$.40/lb.) with approximately 50% due to uilities. A major
difference between the two processes is the relatively
high cost of a plant for the former vs. the latter
(approximately $14.2 vs. $8.4 million for a 6 MCF per year
facility) which could increase the net cost of production
to approximately $.64 vs. $.94 MCF respectively. This
analyses suggests that membranes may be an attractive
choice for high CO_2 content stream processing.

As noted previously, flue gas is the largest potential
source of CO_2. However, recovery costs must be
sufficiently low to allow for acceptable end product
values. The recovery of CO_2 from flue gases is
accomplished with aqueous solutions of MEA. An analysis
of the cost of proprietary technologies for recovery of a
CO_2 generation plant with a CO_2 recovery capacity of 1000
tons/day reveal a net cost of producing CO_2 to be

approximately \$2.75-3.00/MCF with approximately 1/3 allocated to steam costs.

With both processes, 97 percent of the CO_2 is recovered as product. The methane concentration in the CO_2 product is 3 percent. This reflects a 72.5 percent recovery of methane as a by-product. A second stage of membranes would definitely not be required in this case as it would be in the case of feed streams containing less CO_2.

Summary

Although the dilute sources of carbon dioxide make up the greatest volume of available feed stock, the cost of recovery can be a significant component of the overall cost of a reaction process using such CO_2. As membrane separation processes are further developed, they may become the most cost effective route for recovery and utilization of point source generated carbon dioxide.

RECEIVED October 6, 1987

Chapter 2

Carbon Dioxide Equilibria

James N. Butler

Division of Applied Sciences, Harvard University, Cambridge, MA 02138

The solubility of carbon dioxide in aqueous and non-aqueous solutions depends on its partial pressure (via Henry's law), on temperature (according to its enthalpy of solution) and on acid-base reactions within the solution. In aqueous solutions, the equilibria forming HCO_3^- and $CO_3^=$ depend on pH and ionic strength; the presence of metal ions which form insoluble carbonates may also be a factor. Some speculation is made about reactions in nonaqueous solutions, and how thermodynamic data may be applied to reduction of CO_2 to formic acid, formaldehyde, or methanol by heterogenous catalysis, photoreduction, or electrochemical reduction.

The solubility of carbon dioxide in aqueous or nonaqueous media depends on three primary factors: temperature, partial pressure of carbon dioxide, and acid-base reactions in the solution. Accurate data for solubility and equilibria are well-known for aqueous solutions (1-3), but not for nonaqueous solutions. Neither the standard compilations of equilibrium constants (1,2) nor recent reviews on nonaqueous electrolytes (4) cover what appears to be a large and poorly indexed literature.

The reason I say "poorly indexed" is that out of several thousand entries in the 10th collective index to Chemical Abstracts (1977-81), relating to carbon dioxide, only two entries contained the term "solubility" and neither of these pertained to the solubility of carbon dioxide in a liquid phase. On the other hand, the many entries under terms like "removal from natural gas" implied that quite a lot of data could be found with enough effort.

A historical perspective may give a rough estimate: A compilation reporting work done before 1928 (5), contains three and a half large pages of closely-packed quantitative data for the solubility of carbon dioxide in 48 nonaqueous solvents. A 1958 collection (6) gives 17 pages of tables summarizing the solubility of carbon

0097–6156/88/0363–0008$06.00/0

dioxide in 45 organic solvents, as well as salt solutions and mixtures of solvents. If the amount of published information in this
field has doubled every 7 years, as it has in most sciences, the
current literature (after 28 years) should be about 16 times as
large, or about 270 pages of tables.

If a recent critical review of data for carbon dioxide in nonaqueous media has not yet been compiled, it certainly should be
commissioned as soon as possible.

Pressure

The solubility of carbon dioxide under conditions where it does not
undergo significant reaction (such as acidic aqueous solutions) is
governed by Henry's law: the concentration of CO_2 in solution is
proportional to its partial pressure in the gas phase ($\underline{3}$):

$$[CO_2]\gamma_o = K_H^o \, P_{CO_2}$$

Nonideality in the solution phase, resulting from the salting-
out effect (see below), is accounted for by an activity coefficient
γ_o. Nonideality in the gas phase at high pressure can be accounted
for by considering the coefficient K_H to be pressure-dependent, or
by introducing a fugacity coefficient multiplying P, as is common in
the Chemical Engineering literature. The Henry's law constant is
also frequently represented as H, the reciprocal of K_H ($\underline{7}$):

$$y_a \, \Phi_a \, P = m_a \, \gamma_a \, H_a$$

where P is total pressure, Φ_a is fugacity coefficient of CO_2, y_a is
gaseous mole fraction, m_a is molality of CO_2 in the aqueous phase,
and γ_a is the activity coefficient of CO_2 in the aqueous phase.

At ambient pressures and temperatures, Henry's law is essentially linear; but at lower temperatures and higher pressures (par-
ticularly in the supercritical region) nonlinearity in the pressure
dependence of solubility can be quite substantial. In general, K_H
is somewhat higher at higher pressure. For example (data selected
from Ref. $\underline{5}$), K_H in water at $100^\circ C$ and 70 atm is 0.00343; at 140
atm it is 0.00408; K_H in acetone at $-78^\circ C$ and 50 torr (0.0658 atm)
is 12.28; at 650 torr (0.855 atm) it is 13.54.

Temperature

Temperature is the most important parameter affecting the Henry's
law coefficient. Table 1 lists a few randomly (not critically)
selected values for aqueous and nonaqueous media ($\underline{5}$). In water, K_H
decreases by a factor of ten as temperature is raised from $0^\circ C$ to
$200^\circ C$, but above that it increases with increasing temperature.

One empirical equation describing the temperature dependence of
K_H from 0 to $100^\circ C$ is ($\underline{7}$)

$$-\log K_H = \log H = 470.067 - 14947.2/T - 79.163 \ln T + 0.10926 \, T$$

Table I.   Henry's Law Coefficient for CO_2

Temp $^{\circ}C$	Pressure	Solution	K_H	Ref
-78	50 torr	acetone	12.28	ICT
-78	650 torr	acetone	13.54	ICT
-78	740 torr	methanol	7.92	ICT
-78	700 torr	ethanol	4.67	ICT
-65.3	1 atm	99% ethanol	1.76	ICT
0	1 atm	water	.0776	Stumm
25	1 atm	water	.03388	Stumm
25	1 atm	water	.03367	ICT
25	1 atm	water	.03374	ICT
25	1 atm	3 M Aqueous NaCl	.01698	JNB
25	1 atm	3.2 M NH_4Cl	.02717	ICT
20	1 atm	95.6 % H_2SO_4	.0412	ICT
25	1 atm	CCl_4	.0938	ICT
25	1 atm	CS_2	.140	ICT
25	1 atm	$CHCl_3$	.141	ICT
25	1 atm	methanol	.139	ICT
25	1 atm	$C_2H_4Cl_2$	.144	ICT
25	1 atm	99% ethanol	.125	ICT
20	50 atm	ethanol	.157	ICT
25	1 atm	pyridine	.149	ICT
25	1 atm	benzene	.0991	ICT
20	20 atm	benzene	.150	ICT
20	20 atm	chlorobenzene	.130	ICT
25	1 atm	aniline	.0541	ICT
20	1 atm	petroleum	.0522	ICT
100	1 atm	water	.01023	Stumm
100	70 atm	water	.00343	ICT
100	140 atm	water	.00408	ICT
100	50 atm	ethanol	.02936	ICT
100	135 atm	ethanol	.04363	ICT
100	97 atm	ethyl ether	.05842	ICT
200	16 atm	water	.00891	Stumm
330	90 atm	water	.0200	Ellis

Notes: Temperature has the biggest effect.
K_H increases at lower T.
Pressure has small effect: K_H increases at higher P.
K_H is larger in organic solvents than in water.
No dramatic difference among solvents.

Refs: Stumm, W.; Morgan, J.J. Aquatic Chemistry; New York: Wiley,
2nd Ed., 1981. See Ref. (3) in text.
Ellis, A.J. Golding, R.M. Am. J. Sci. 1963, 261, 47-60.
ICT = International Critical Tables (1928). Ref. (5) in
text.

where T is in degrees Kelvin. At 25°C, log K_H = -1.472, or K_H = 0.0337, in agreement with Table 1. An alternative empirical equation (8) covers the range from 0 to 250°C:

$$-\log K_H = 41.0371 - 2948.44/T - 4.9734 \; \ln T - 0.0045401 \; T$$

At 25°C this gives log K_H = -1.458, or K_H = 0.0348, a little higher than the data in Table I. However, the wider temperature range is an advantage for some purposes.

At low temperatures carbon dioxide is extremely soluble in most polar organic solvents such as alcohols and ketones (Table I).

Solvent

At ambient temperature, carbon dioxide is three to five times more soluble in most organic solvents than in water (Table I). The differences among polar (e.g. methanol, K_H = 0.139) and nonpolar (e.g. carbon tetrachloride, K_H = 0.094) solvents are small. Two solvents which have recently been of practical interest in removing carbon dioxide from natural gas are propylene carbonate (9) and monoethanolamine (10) - this last ought to be classified as an acid-base reaction. Judging from the number of entries in the 10th collective index to Chemical Abstracts, there is a substantial chemical engineering literature on this topic.

Salts

The ions of nonreactive salts influence the solubility of CO_2 via the salting-out effect. This is accounted for by an activity coefficient for uncharged CO_2, whose logarithm is directly proportional to the ionic strength. At 25°C (3),

$$\log \gamma_o = 0.1 \; I$$

for sodium chloride solutions. At other temperatures and higher ionic strengths, this empirical equation (11, 12) can be used:

$$\log \gamma_o = \frac{(33.5-0.109t+0.0014t^2)I - (1.5+0.015t+0.004t^2)I^2}{(t + 273)}$$

where t is temperature in degrees Celsius.

Although the salt effect is somewhat greater at higher temperatures, it is not large compared to the effects of pressure, temperature, solvent, and especially acid-base reactions.

Acid-Base Reactions

Reaction of CO_2 with bases - either as solvent or solute - is by far the most significant effect on CO_2 solubility in aqueous media. The equilibria are well-known for aqueous solutions (1, 2, 3, 7), but little data has been systematically compiled on acid-base reactions of CO_2 in nonaqueous solutions (see Introduction).

Effects of other ions in solution on the acid-base reactions are accounted for in two complementary ways - via activity coefficients in the equilibrium equations (3):

$$[H^+][HCO_3^-] \, \gamma_+ \, \gamma_- = K_{a1} \, [CO_2] \, \gamma_o$$
$$[H^+][CO_3^=] \, \gamma_+ \, \gamma_= = K_{a2} \, [HCO_3^-] \, \gamma_-$$

and via equilibria of ion-pairing, exemplified by the introduction of a sodium-carbonate ion pair equilibrium

$$[NaCO_3^-] = K_i \, [Na^+][CO_3^=] \; .$$

The ion pair is included in addition to Na^+, HCO_3^-, and $CO_3^=$ in the mass and charge balances (3):

$$C_T = [CO_2] + [HCO_3^-] + [CO_3^=] + [NaCO_3^-]$$
$$[H^+] + [Na^+] = [HCO_3^-] + 2 \, [CO_3^=] + [NaCO_3^-] + [OH^-]$$

The distinction between "pure" electrostatic effects and ion association is somewhat arbitrary; for example, many workers have used an extended Debye-Huckel equation with fixed ion-size parameters to compute the activity coefficients, and assigned all other nonideality to ion-pairing equilibria (3, 13, 14). Recent reviews have surveyed more elaborate solvation-based thermodynamic theories of electrolytes (15, 16).

The first acidity constant decreases monotonically with temperature: $pK_{a1} = -\log (K_{a1})$ changes from 6.58 at 0^oC to 6.352 at 25^oC to 6.30 at 65^oC (7). An empirical relation for the range $0-65^oC$ is:

$$\log K_{a1} = 1202.09 - 34771.1/T - 207.876 \, \ell n \, T + 0.310514 \, T$$

pK_{a1} increases to 7.08 at 200^oC and 8.3 at 300^oC (3, 8, 17). An empirical relationship covering the range from $0-225^oC$ is (8):

$$\log K_{a1} = 102.268 - 5251.53/T - 15.9740 \, \ell n \, T$$

which gives $pK_{a1} = 6.36$ at 25^oC and 7.22 at 200^oC.

An extensive dissertation on the prediction of thermodynamic properties for aqueous electrolytes can be found in the work of Helgeson and Kirkham (18).

The second acidity constant also goes through an extremum: pK_{a2} decreases from 10.625 at 0^oC to a minimum of 10.14 at about 90^oC, then increases to 10.89 at 218^oC (17).

At high pressures and temperatures, or in nonaqueous media, pH may not be a convenient quantity to measure; however, any quantity can be made into a master variable via the appropriate mass and charge balances. One obvious quantity is the partial pressure of CO_2; another is the titration alkalinity (3)

$$A_c = [HCO_3^-] + 2 [CO_3^=] + [OH^-] - [H^+] + \dots$$

which to a first approximation is independent of CO_2 concentration.

In the absence of a good literature survey, let me speculate about what effects one might expect in nonaqueous media. In the total absence of water, CO_2 would not form H_2CO_3, and hence would not provide HCO_3^- or $CO_3^=$. On the other hand, a strongly basic solvent or solute might form analogous ions with CO_2. For example:

$$CO_2 + OH^- = HCO_3^- ;$$

$$CO_2 + NH_2^- = CO_2NH_2^- .$$

In solvents of low dielectric constant (i.e. carbon tetrachloride or benzene), dissociation into ions is more difficult, and ion association is the rule rather than the exception.

In aprotic solvents of high dielectric constant (i.e. propylene carbonate or dimethyl sulfoxide), dissociation into ions is facilitated, but to produce HCO_3^- one would have to provide some water to react with CO_2. Hydroxides and carbonates tend to be insoluble in anhydrous aprotic solvents (19), and thus HCO_3^- and $CO_3^=$ would tend to be present at very low concentration, if at all.

Electrochemistry

Reduction of carbon dioxide can produce a wide variety of possible products. Thermodynamically, the most stable product is methane, but products of intermediate oxidation state such as methanol, methanal, formate, oxalate, carbon monoxide, and elemental carbon are all possibilities (20, 21).

CO_2 reduction proceeds readily to formic acid on most metal electrodes, and formic acid reduction proceeds most rapidly on electrodes with high hydrogen overvoltage such as lead, tin, and indium; this appears to be related to the stability of intermediates (22, 23).

The anions HCO_3^- and $CO_3^=$ tend to be repelled from the negatively charged electrode. It has been established (24) that even in concentrated solutions of $NaHCO_3$, reduction of bicarbonate ion occurs via uncharged CO_2. Direct reduction of the anion is very slow by comparison.

Reduction of carbon dioxide and its intermediates is catalyzed by illumination. Photocatalyzed reduction of CO_2 to formaldehyde and methanol has been observed on semiconductor powders suspended in solution (25) as well as on GaAs electrodes (26).

One might expect that the higher concentrations of CO_2 in solution, obtained in nonaqueous media, might increase the rate of reduction, but since the enhancement of solubility is not great (Table I) and the effect of the different solvent on reaction intermediates is difficult to predict, increasing the rate of CO_2 reduction by changing solvent is probably not a simple or straightforward problem.

Acknowledgments

Preparation of this paper was supported in part by Harvard
University. The author thanks Jeanne Sattely for helping with the
illustrations and transparencies; and Carrie Kent of the Cabot
Science Library for her brief but enthusiastic attack on the
ignorance explosion.

Literature Cited

1. Sillen, L.G.; Martell, A.E. Stability Constants; London: The
   Chemical Society, Special Pub. 17, 1964 and 25, 1971.
2. Martell, A.E.; Smith, R.M. Critical Stability Constants; New
   York: Plenum Press, 4 volumes, 1974-1977.
3. Butler, J.N. Carbon Dioxide Equilibria; Reading, MA: Addison-
   Wesley Publishing Co., 1982.
4. Janz, G.J.; Tomkins, R.P.T.; et al. Nonaqueous Electrolytes
   Handbook; New York: Academic Press, 2 volumes, 1972-1973.
5. Washburn, E.W. et al. International Critical Tables;
   Washington, DC: National Research Council, 1928; Vol. 3, pp.
   260-283.
6. Seidell, A.; Linke, W.F. Solubilities; New York: Van Nostrand,
   4th ed., 1958; Vol. 1, pp. 476-493.
7. Edwards, T.J.; Newman, J.; Prausnitz, J.M. "Thermodynamics of
   aqueous solutions containing volatile weak electrolytes,"
   A.I.Ch.E. Journal 1975, 21(2), 248-259.
8. Edwards, T.J.; Maurer; G., Newman, J; Prausnitz, J.M. "Vapor-
   liquid equilibria in multicomponent aqueous solutions of
   volatile weak electrolytes," A.I.Ch.E. Journal 1978, 24,
   966-976.
9. Freireich, E.; Tennyson, R.N. Proc. Gas. Cond. Conf. 1977, 27,
   D/1-D/11, Chem Abstr. 1977, 88, 107620W.
10. DeCoursey, W.J. Des. Process Plants. Can. Chem. Eng. Conf. 27th
    1977, 136-142, Chem. Abstr. 1977, 88, 109792C.
11. Wigley, T.M.L.; Plummer, L.N. Geochim. Cosmochim. Acta 1976,
    40, 989-995.
12. Harned, H.S.; Davis, R. Jr. J. Am. Chem. Soc. 1943, 65, 2030-
    2037.
13. Garrels, R.M.; Thompson, M.E. Am. J. Sci. 1962, 260, 57-66.
14. Whitfield, M. Limnol. Oceanogr. 1974, 19, 235-245.
15. Maurer, G. Fluid Phase Equil. 1983, 13, 269-296.
16. Horvath, A.L. Handbook of Aqueous Electrolyte Solutions; New
    York: Ellis Horwood Ltd. - John Wiley & Sons, 1985, pp.
    206-232.
17. Helgeson, H.C. "Thermodynamics of complex dissociation in aque-
    ous solution at elevated temperatures," J. Phys. Chem. 1967,
    71, 3121-3136.
18. Helgeson, H.C.; Kirkham, D. "Theoretical prediction of the
    thermodynamic behavior of aqueous electrolytes at high temper-
    atures and pressures," Part I: Am. J. Sci. 1974, 274, 1089-
    1198; Part II: Am. J. Sci. 1974, 1199-1261, Part III: Am. J.
    Sci. 1976, 97-240.
19. Butler, J.N. J. Electroanal. Chem. 1967, 14, 89-116.

20. Pourbaix, M. Atlas of Electrochemical Equilibria; Brussels: CEBELCOR and Houston, TX: National Assoc. of Corrosion Engineers, 1958, pp. 449–457.
21. Bard, A.J.; Parsons, R.; Jordan, J., Eds. Standard Potentials in Aqueous Solution; New York: Marcel Dekker, 1985, pp. 189–200.
22. Russell, P.G., et al. J. Electrochem. Soc. 1977, 124, 1329–1338.
23. Kapusta, S.; Hackerman, N. J. Electrochem. Soc. 1983, 130, 607–613. This paper contains a brief review of the earlier literature.
24. Hori, Y.; Suzuki, S. J. Electrochem. Soc. 1983, 130, 2387–2390.
25. Inoue, T., et al. Nature 1979, 277, 637–638.
26. Frese, K.W.; Canfield, D. J. Electrochem. Soc. 1984, 131, 2518–2522.

RECEIVED June 24, 1987

Chapter 3

Coordination of Carbon Dioxide to Nickel

An Alternative Theoretical Model

R. P. Messmer and H.-J. Freund[1]

General Electric Company, Corporate Research and Development,
Schenectady, NY 12301

An alternative to the molecular orbital description of CO_2 bonding to a transition metal is proposed here. The new description is based on ab initio calculations which include important electronic correlation effects neglected in molecular orbital theory. The resulting valence bond picture, which includes "bent-bonds" for CO_2 rather than σ and π bonds, has striking similarities to the description given in qualitative discussions by Pauling many years ago.

In the last two to three decades molecular orbital theory has become the paradigm for discussing bonding in molecules. It has had many impressive successes and has contributed greatly to our understanding of the electronic structure of molecules. However, one must not lose sight of the fact that molecular orbital theory totally neglects electronic correlation effects, which may have important consequences for bonding.

Here, we investigate the effect of electronic correlation on our understanding of chemical bonding for the case of the CO_2 molecule and the coordination of this molecule to a nickel atom. We employ ab initio calculations based on the generalized valence bond (GVB) method (1) to study the CO_2 molecule, both as an isolated entity and coordinated to Ni. By analogy to transition metal complexes, three different coordination geometries for the CO_2 molecule are considered: pure carbon coordination (I), pure oxygen coordination (II) and mixed carbon-oxygen coordination (III).

[1]Current address: Institut für Physikalische und Theoretische Chemie der Friedrich-Alexander-Universität Erlangen-Nürnberg, Egerlandstrasse 3, D–8520 Erlangen, Federal Republic of Germany

0097–6156/88/0363–0016$06.00/0

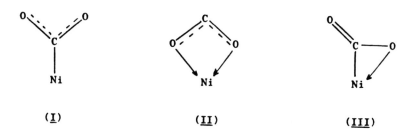

(**I**)   (**II**)   (**III**)

In all cases studied, the bonding between the CO_2 moiety and the metal atom is described best as a CO_2^- anion interacting with a Ni cation. On the basis of the calculations, several features of the interaction of CO_2 and metal surfaces can be discussed. Before discussing these results for coordinated CO_2 in more detail, we consider the isolated CO_2 molecule.

The CO_2 Molecule and Bent-Bonds

It has recently been demonstrated (2) that if correlated wave functions are used in the description of the CO_2 molecule and the orbitals are not <u>forced</u> to be symmetry orbitals, then one obtains a lower energy for multiple "bent-bonds" (Ω-bonds) than for the traditional σ and π bonds. Figure 1 shows contour plots of the Ω-bonds for one of the double bonds of CO_2. In panel A two orbitals which overlap to form a C-O bond are shown. On the left is an orbital more localized on the oxygen atom and on the right one which is more localized on the carbon atom; they can be thought of as overlapping, variationally determined atomic-like hybrids. Panel B shows the orbitals which make up the other half of the double bond; they are clearly symmetry-related to those in panel A. In panel C, contour plots of one of the oxygen lone pairs are shown. One can clearly see the "in-out" correlation exhibited by the pair of electrons, in which one orbital (at the right) is closer to the oxygen nucleus, while the other orbital (at the left) is more extended. Panel D shows the orbitals making up the second lone pair on the same oxygen atom; they are equivalent by symmetry to the pair in panel C. In Figure 2a, a schematic representation of the many-electron wave function is presented. The dots denote the number of electrons in each orbital and the lines denote which orbitals overlap to form bonds. The bonds labeled A and B in Figure 2a are composed of the orbitals in panels A and B of Figure 1, respectively. The computational details are described elsewhere (2).

The perfect-pairing (PP) orbitals of this wave function clearly show the "lone-pairs" and "bond pairs" which are part of the language of the experimental chemist. This is in contrast to the molecular orbital description or to the GVB description with $\sigma - \pi$ restrictions where the lone pairs and "π" bonds are not discernable from contour plots of the orbitals (2). It is somewhat reassuring that the wave function which gives the lowest variational energy (that of Figures 1 and 2a) also most closely coincides with the experimental chemist's traditional view of the bonding (3).

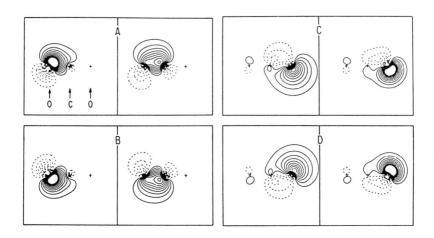

Figure 1. GVB-perfect pairing orbitals for the ground state of CO_2:
A and B show the orbitals of the Ω-bonds for one of the CO double
bonds; C and D show the orbitals of the two lone pairs on one of the
oxygen atoms.

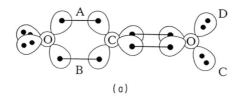

(a)

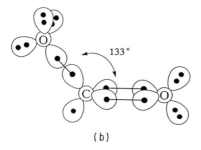

(b)

Figure 2.   Schematic representations of the many-electron wave
functions for: (a) the CO_2 ground state; (b) the CO_2^- anion.

The CO_2^- anion is unstable in the gas phase, but the dimer anion $(CO_2)_2^-$ which can be thought of as composed of a neutral molecule (somewhat polarized of course) and a carbon dioxide anion, is stable (<u>4</u>). We have investigated the CO_2^- anion using this geometry and find the many-electron wave function to be of the form shown schematically in Figure 2b. The lone pair and bond pair orbitals of CO_2^- are found to be nearly identical to those of CO_2 (Figure 1), supporting the intuitive notion of transferability of such entities between similar systems. One can think of the wave function of CO_2^- (Figure 2b) as arising from that of CO_2, when an additional electron is accepted by the latter in an orbital of an oxygen atom (the oxygen orbital of bond B, for example). The extra electron on oxygen (due to its higher electron affinity) results in the breaking of a carbon-oxygen bond and the formation of another oxygen lone pair. The anion can lower its energy by allowing the remaining bond (A in Figure 2a) to increase the overlap of its component orbitals (and decrease the Pauli repulsion between the new lone pair and the singly occupied orbital on carbon), resulting in the wave function of Figure 2b. This provides a simple and natural alternative explanation for the geometry of CO_2^- to that given by Walsh's rules.

Coordinated CO_2

We now turn to a discussion of the bonding between CO_2 and a Ni atom. A complete description of theoretical and computational details and the relationship of the results to the chemisorption of CO_2 on metal surfaces has been presented elsewhere (<u>5</u>); here we focus strictly on those aspects related to the bonding. However, before we discuss the bonding for the three individual coordination geometries (<u>I</u>), (<u>II</u>) and (<u>III</u>) on the basis of the PP-orbitals, it is appropriate to comment on a general result which is independent of the geometry. That is, all the orbitals localized on the CO_2 moiety in the $NiCO_2$ complexes have the same shapes as the orbitals of CO_2 with the exception of those orbitals directly interacting with the Ni atom which are modified as described in the following discussion. This general result suggests that it is indeed appropriate to consider the bonding in Ni-CO_2 by focusing on the interaction of CO_2^- and Ni^+.

Pure Carbon Coordination (I)

For the case of the pure carbon coordination (<u>I</u>), it is obvious that the single electron on the CO_2^- forms a covalent bond with the unpaired d-electron on the Ni atom in its d^9 configuration. The GVB pair forming this bond is shown in Figure 3a. Although the results of a Mulliken population analysis can only be regarded as a qualitative indicator of the actual charge distribution, a calculated electron transfer of 0.53e from the Ni atom to the CO_2 moiety supports the view expressed above regarding the substantial ionic nature of the interacting species. Figure 3b shows a schematic representation of the many electron wave function; the bond whose orbitals are given in Figure 3a, is labeled A. The stabilization of (<u>I</u>) with respect to the infinitely separated parts is 7.79 eV. This stabilization is due both to the formation of a covalent bond and the coulombic interaction. An upper

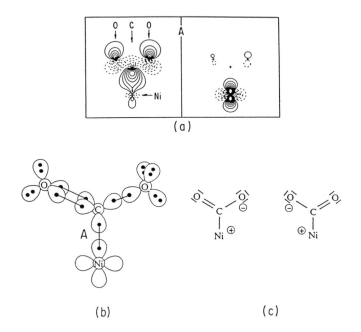

Figure 3. Coordination of CO_2 to Ni via carbon coordination: (a) orbitals forming the Ni–C covalent bond; (b) schematic diagram of the many-electron wave function; (c) classical "resonance" structures.

limit to the latter contribution can be estimated from the Mulliken atomic charges of the separated systems to be 5.65 eV, which is 73% of the total stabilization energy. Note that the corresponding energy using the Hartree-Fock (molecular orbital) wave function is only 5.68 eV, i.e., close to the pure coulomb interaction energy. However, the proper reference for the total energy of the system is not $Ni^+ + CO_2^-$ but rather $Ni + CO_2$. With respect to this reference point (I) is unbound by 1.72 eV. At the Hartree-Fock level it is unbound by 3.57 eV. As discussed elsewhere (5), this analysis with respect to the separated neutral species using calculated total energies needs to include corrections for errors made in the calculated electron affinity of CO_2 and the calculated ionization potential of the Ni atom. To correct for these errors we need to add 0.86 eV (HF: 1.02 eV) to the calculated binding energy with respect to the neutral species. Even including this correction, however, leaves (I) unbound by 0.86 eV (HF: 2.55 eV) with respect to $Ni + CO_2$. Although two resonance structures (Figure 3c) should be taken into account for a proper description of the ground state wave function, the resonance stabilization will mainly arise on the CO_2^- moiety, and therefore be similar to that in uncoordinate CO_2^-. Thus we expect only little influence of resonance on the bond energy of (I). Therefore, we conclude that the pure carbon coordination is unfavorable for CO_2 bonding to Ni.

Pure Oxygen Coordination (II)

If we place the Ni atom on the opposite side of the CO_2 moiety as compared to (I), the Ni atom has a pure oxygen coordination, (II). We have chosen the Ni-O distance to be consistent with the bond lengths found in molecular complexes (6,7). The panels (A, A', B and C) of Figure 4a show the orbitals which are non-trivially modified as a result of the Ni-CO_2 bonding interaction. The system has two unpaired electrons, coupled to form a triplet state. One electron resides on the CO_2 moiety pointing away from the Ni atom (panel A), and the other is a d-electron on the Ni atom (panel A'). The former has the same shape as the orbital of the unpaired electron in free CO_2^-. The two lone pairs which establish two dative bonds to the Ni atom are shown in panels B and C of Figure 4a. The bonds are formed by the lone pairs donating into somewhat diffuse, unoccupied hybrid orbitals (of s and p character) on the Ni atom. Due to the choice of the contours in the plotted orbitals (increment of 0.05 a.u.) the bonding interaction between the oxygen lone pairs and the diffuse Ni orbitals shows up as an indentation in the contours of the in-out correlated lone pairs which is not present in the isolated system (Figure 1). A schematic representation of the wave function is given in Figure 4b. A charge transfer similar to that for (I) is found, however the stability of (II) has drastically improved over that of (I). With respect to the separated ions, namely Ni^+ and CO_2^-, we calculate a stabilization energy of 9.23 eV. With respect to the separated neutrals, (II) is unbound by 0.28 eV. This results in a bound state by 0.58 eV, once the correction for the electron affinity of CO_2 and the ionization potential of Ni are taken into account.

The Hartree-Fock calculation for (II) yields a stabilization energy with respect to the separated atoms which is nearly identical to the GVB-PP calculation, and taking the appropriate correction into

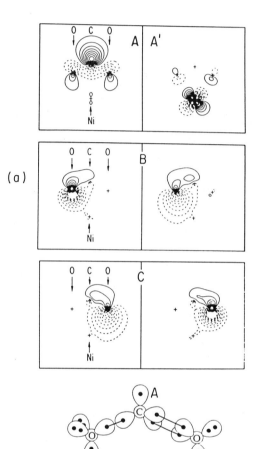

Figure 4. Coordination of CO_2 to Ni via pure oxygen coordination:
(a) triplet coupled orbitals (A and A'); oxygen lone pairs forming
dative bonds to Ni (panels B and C); (b) schematic diagram of the
many-electron wave function.

account (see above) the system is stable with respect to dissociation into CO_2 and Ni. Thus, as far as the calculated stabilization energy is concerned, electron correlation does not seem to be crucial in this case, because the amount of covalent bonding is smallest, and the electrostatic interaction largest, of the three geometries considered. The extra coulombic stabilization experienced in (II) as compared to the pure carbon coordination (I) is sufficient to account for the higher stability of (II). Thus, our calculations indicate that the pure oxygen coordination represents a favorable coordination geometry for the Ni-CO_2 interaction. It should be mentioned that Jordan (8) has predicted a corresponding structure for $LiCO_2$ on the basis of Hartree-Fock calculations. The structure of the Li-salts have recently been investigated using vibrational spectroscopy (9) and the results seem to be consistent with Jordan's prediction.

Mixed Carbon-Oxygen Coordination (III)

The final coordination geometry considered is (III). We have chosen the geometry such that the carbon-Ni and oxygen-Ni bonds lengths are consistent with those of (I) and (II). Figure 5a shows the orbitals involved in the bonding between Ni and CO_2. In view of the discussion above, it is almost unnecessary to note that all other orbitals are basically identical to the non-interacting fragments Ni^+ and CO_2^-. Panel A of Figure 5a shows the GVB pair represening the carbon-Ni bond. Except for the asymmetry induced by the chosen geometry, the bond is very similar to the one shown in panel A of Figure 3a for the case of pure carbon coordination. Panel B shows the oxygen lone pair donation into the diffuse s/p hybrid orbitals of Ni as indicated by the indentation of the lone pair contours. Clearly, this unsymmetric coordination involves both covalent and dative bonding modes. A schematic view of the many-electron wave function is shown in Figure 5b. The stability of this coordination mode is nearly as great as the pure oxygen coordination (II). Similar to the situation for (I), the present geometry is unbound for the uncorrelated Hartree-Fock calculations even after accounting for the appropriate corrections. Thus, correlation effects for (III) are crucial in obtaining a bound system, unlike the situation for (II).

Summary

Our results for the three coordination modes considered here indicate that the CO_2 molecule prefers to adopt either a mixed carbon-oxygen coordination (III) or a pure coordination (II) to the metal center, while a pure carbon coordination (I) appears to be unfavorable. Further, the results suggest a rather weak CO_2 - transition metal bond consistent with the low stability of adsorbed CO_2 and CO_2 in molecular complexes.

We have attempted to summarize some qualitative aspects of bonding obtained from recent quantitative calculations which include important electronic correlation effects ignored in molecular orbital theory. The resulting valence bond concepts derived from the calculations, which have long been ignored as only qualitative and without sound theoretical foundation, are made quantitative and computationally accessible through the generalized valence bond theory. The concept of bent-bonds (Ω-bonds), much discussed in the chemical

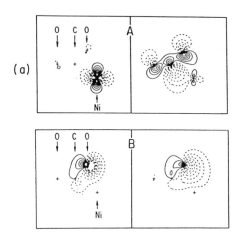

(a)

(b)

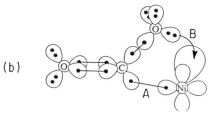

Figure 5. Coordination of CO_2 to Ni via mixed carbon and oxygen coordination: (a) orbitals forming a covalent bond between Ni and C (panel A); orbitals of a oxygen lone pair forming a dative bond to Ni (panel B); (b) schematic diagram of the many-electron wave function.

literature for decades has been demonatrated for the first time (2) to be the energetically favored description of bonding in CO_2. The Ω-bonds description is not only found for double bonds, but for triple bonds and conjugated bonds as well (Messmer, R. P. and Schultz, P. A. Physical Review Letters, in press). Hence, it appears that a natural language for the discussion of electronic correlation effects in bonding is a modified version of the valence bond approach long used by chemists as a useful empirical method to organize their vast experimental experience.

Literature Cited

1. Bair, R. A.; Goddard, W. A., III; Voter, A. F.; Rappé, A. K.; Yaffe, L. G.; Bobrowicz, F. W.; Wadt, W. R.; Hay, P. J.; Hunt, W. J. GVB2P5 (unpublished); see Bair, R. A. PhD Thesis, Caltech (1980); Hunt, W. J.; Hay, P. J.; Goddard, III, W. A. J. Chem. Phys. 1972, 57, 738; Bobrowicz, F. W.; Goddard, W. A., III in Modern Theoretical Chemistry: Methods of Electronic Structure Theory; Schaefer, H. F., Ed.; Plenum: New York, 1977; Vol. 3, p 79.
2. Messmer, R. P.; Schultz, P. A.; Tatar, R. C.; Freund, H.-J. Chem. Phys. Lett. 1986, 126, 176.
3. Pauling, L. The Nature of the Chemical Bond; Cornell University Press, New York, 1960; 3rd Edition.
4. Bowen, K. H.; Liesegang, G. W.; Sanders, R. A.; Herschbach, D. R. J. Phys. Chem. 1983, 87, 557.
5. Freund, H.-J.; Messmer, R. P. Surface Sci. 1986, 172, 1.
6. Aresta, M.; Nobile, C. F. J. Chem. Soc. Dalton Trans. 1977, 708.
7. Beck, W.; Raab, K.; Nagel, U.; Steinmann, M. Angew. Chem. 1982, 94, 526.
8. Jordan, K. D. J. Phys. Chem. 1984, 88, 2459.
9. Kafafi, Z. H.; Hange, R. H.; Billups, W. E.; Margrave, J. L. J. Am. Chem. Soc. 1983, 105, 3886.

RECEIVED March 3, 1987

Chapter 4

Metal-Induced Transformations of Carbon Dioxide

Donald J. Darensbourg, Christopher G. Bauch, and Cesar Ovalles

Department of Chemistry, Texas A&M University, College Station, TX 77843

The activation of carbon dioxide by homogeneous and heterogeneous metal catalysts, as well as the nature of the stoichiometric insertion processes of these catalysts, are examined. The kinetic and mechanistic aspects of CO_2 insertion into the M-X bond of $M(CO)_5X^-$ complexes (M = W, Cr and X = H, alkyl, aryl, aryloxy, and alkoxy) is investigated. The mechanism of CO_2 insertion in these systems is described as an associative interchange (I_a) type mechanism where prior loss of coordinated CO is not involved in the insertion process. The homogeneous catalytic transformations of CO_2 involve the formation of alkyl formates from alcohols and alkyl halides using the anionic tungsten complexes, $W(CO)_5Y^-$ (Y = $-OOCCH_3^-$, $-\mu-H-W(CO)_5^-$, and $-Cl^-$), as catalysts. Alumina supported ruthenium clusters were studied for the effect of cluster nuclearity on the rate of CO_2 methanation. It was found that the reactivity paralleled the nuclearity of the catalyst precursor.

The chemistry of one-carbon molecules (better known as C_1 chemistry) is a very exciting area of research for the organometallic chemist. The motivation for these efforts stems from the belief that the raw material base for commercial organic chemicals will shift from oil to coal, in the near future, due to the decline of petroleum reserves. Of the raw materials for the C_1-based industry, carbon monoxide is the most commonly used and a great deal of the current research effort is designed to investigate the activation of this molecule (1–5).

An alternative source of chemical carbon is carbon dioxide, which is the cheapest and most abundant of the C_1 molecules (6–12). This single-carbon species has been widely neglected mainly because it is regarded as a highly stable molecule. For example, it is the thermodynamic end-product of many energy producing processes, the most prominent being the combustion of hydrocarbons. Nevertheless, there are many thermodynamically favorable reactions of CO_2 which provide useful organic substances (Equations 1-3). The standard

0097–6156/88/0363–0026$06.00/0

enthalpy changes associated with the corresponding reactions involving carbon monoxide are listed for comparison.

$$CO_2(g) + H_2(g) \longrightarrow HCOOH(\ell) \qquad -15.7 \text{ kJ/mol}$$
$$CO(g) + H_2O(\ell) \longrightarrow HCOOH(\ell) \qquad -12.9 \text{ kJ/mol}$$

(1)

$$CO_2(g) + 3H_2(g) \longrightarrow CH_3OH(\ell) + H_2O(\ell) \qquad -130.9 \text{ kJ/mol}$$
$$CO(g) + 2H_2(g) \longrightarrow CH_3OH(\ell) \qquad -128.1 \text{ kJ/mol}$$

(2)

$$CO_2(g) + H_2(g) + CH_3OH(\ell) \longrightarrow HCOOCH_3(\ell) + H_2O(\ell) \qquad -32.2 \text{ kJ/mol}$$
$$CO(g) + CH_3OH(\ell) \longrightarrow HCOOCH_3(\ell) \qquad -23.3 \text{ kJ/mol}$$

(3)

The two processes are of course related by the water-gas shift reaction (Equation 4). Although carbon dioxide is more stable than carbon monoxide by 283 kJ/mol, formation of the very stable water

$$CO(g) + H_2O(\ell) \longrightarrow CO_2(g) + H_2(g) \qquad +2.8 \text{ kJ/mol} \qquad (4)$$

molecule in processes involving the former C_1 species accounts for the thermodynamic similarity of these reactions. Indeed it is the requirement of an extra mole of hydrogen in the CO_2 reactions for water production which makes it generally the less attractive process.

In 1985, over 4 million tons of carbon dioxide were produced from non-oilfield sources (13). About three-fourths of this carbon dioxide is produced as a co-product in the manufacture of ammonia. Recovery as a co-product from grain fermentation provides another major source of carbon dioxide. It is also obtained from refinery and chemical operations and natural wells. The major commercial uses of carbon dioxide are derived from its physical properties. These uses include refrigeration, beverage carbonation and fire extinguishers. Only 10% of the carbon dioxide produced is used in chemical manufacture.

Currently, carbon dioxide is used as a chemical feedstock for the production of carboxylic acids, carbonates, carbon monoxide, and urea (14–16). Despite the fact that numerous chemical reactions utilizing carbon dioxide are thermodynamically advantageous, there is often a substantial kinetic barrier to their occurrence. Transition metal compounds can serve to catalyze reactions of carbon dioxide, i.e., in the utilization of carbon dioxide in synthetic organic chemistry, transition metal complexes can simultaneously activate both carbon dioxide and other substrate molecules such as hydrogen or olefins. We have initiated investigations intended to characterize homogeneous carbon dioxide reduction processes and our results to date are summarized herein. Specifically, our research centers on mechanistic studies of metal carbonyl anion catalysts, which readily activate CO_2 to C-H and C-C bond formation, two of the most important processes in the synthesis of organic materials.

INSERTION REACTIONS OF CARBON DIOXIDE

A primary concern in the investigation of CO_2 activation catalysis is an examination of the stoichiometric reactions this molecule undergoes with transition metal complexes. The most important of these reactions are the insertions of CO_2 into metal-hydrogen, -carbon, and -oxygen bonds, because these often represent the first steps in the conversion of CO_2 into organic compounds.

Metal Hydrides. Insertion of CO_2 into the metal-hydrogen bond of cis-HM(CO)$_4$L$^-$ (M = W, Cr; L = CO, PMe$_3$, P(OMe)$_3$) has been found to be an extremely facile process (17–19). This is in contrast with the inability of CO_2 to insert into the metal-hydrogen bond of the analogous neutral manganese hydrides. Although the group 6 and group 7 hydrides are isoelectronic, they have rather different properties. The hydride in HMn(CO)$_5$ is in fact rather acidic with a pKa of ∼7. In contrast, HCr(CO)$_5$$^-$ is very hydridic with a great deal of electron density located at the hydride ligand itself as indicated by ion pairing studies (20). This electron density at the hydride is important in its interaction with the electrophilic carbon of CO_2. By providing a highly localized negative charge, the anionic hydride attracts the carbon dioxide in close to the metal center promoting the orbital overlaps necessary for the formation of the insertion product.

    Another anionic hydride which also undergoes CO_2 insertion is the cluster complex HRu$_3$(CO)$_{11}$$^-$ which gives the bridging formate complex HCO$_2$Ru$_3$(CO)$_{10}$$^-$ (21). The insertion reaction for the cluster hydride is not nearly as facile as that of the group 6 monomers. The ruthenium complex requires high temperatures and pressures of CO_2 in order for insertion to occur. The reason for the difficulty may stem from the delocalization of the negative charge between three metal atoms which would not allow for as strong an interaction with CO_2 as is the case for the monomers.

Metal Alkyls and Aryls. The insertion of CO_2 into metal-carbon bonds allows for the formation of carbon-carbon bonds and is an important step in its activation. In an effort to further define the nature of the CO_2 insertion process, we have examined its occurrence in the anionic group 6 complexes, cis-RM(CO)$_4$L$^-$ (R = —CH$_3$, —C$_2$H$_5$, —C$_6$H$_5$, —CH$_2$C$_6$H$_5$; M = W, Cr; L = CO, PMe$_3$, P(OMe)$_3$) (22,23). Carbon dioxide was found to insert smoothly into the metal carbon bond to form the corresponding carboxylate complexes (Equation 5). The identity of these carboxylates was confirmed by comparison with an authentic

$$[M]-R^- + CO_2 \longrightarrow [M]-O_2CR^- \qquad (5)$$

sample prepared by the reaction of the metal chloride with the silver salt of the carboxylic acid.

    Investigations into the kinetics of CO_2 insertion have revealed that the reaction is first order in both metal substrate and CO_2 (Equation 6). Figure 1 illustrates the linear dependence of the

$$\text{rate} = k_2[CH_3W(CO)_5{}^-][CO_2] = k_{obs}[CH_3W(CO)_5{}^-] \qquad (6)$$

pseudo-first-order rate constant (k_{obs}) on CO_2 pressure for the carboxylation of $CH_3W(CO)_5^-$. At pressures above 200 psi however, the concentration of CO_2 increases more rapidly with further increase in pressure due to a breakdown in Henry's law at elevated pressures. Several factors effect the rate of CO_2 insertion into metal carbon bonds. The most notable of these is the electron density at the metal center. As is the case for the hydride (**vide supra**), the isoelectronic neutral manganese and rhenium analogues of the group 6 alkyls do not undergo CO_2 insertion. Demonstration of the dramatic effect that the electron density, at the metal center, has on CO_2 insertion can be seen when the electron donating property of the ligand, L, in **cis**–$CH_3W(CO)_4L^-$, is varied. Table I shows that the second order rate constant increases by two orders of magnitude when CO is replaced by the better electron donating ligands $P(OMe)_3$ and PMe_3. As expected, the more basic PMe_3 ligand has a greater effect than that of $P(OMe)_3$.

Table I. Second-order rate constants for carbon dioxide insertion into **cis**-$CH_3W(CO)_4L^-$ derivatives.[a]

L	k_2 x 10^4 (Sec^{-1}-M^{-1})	Relative Rates
CO	3.46 x 10^{-2}	1.00
$P(OMe)_3$	2.00	57.8
PMe_3	8.40	243

[a]Reactions carried out in tetrahydrofuran at 25°C at a carbon dioxide pressure of 760 torr.

Although the rate of CO_2 insertion into the metal carbon bond is over 200 times faster than the parent carbonyl, X–ray crystal structures have determined (22)(Darensbourg D. J.; Bauch, C. G.; Rheingold, A. L. Inorg. Chem., in press) that the M–C bond distance is somewhat shorter for the phosphine substituted complex (Table II). This indicates that the strength of the metal carbon bond, as evinced by the M–C bond distance, is of secondary importance compared to the electron density at the metal center. This is plainly evident when comparing the neutral $CH_3Re(CO)_5$ with the anionic $CH_3W(CO)_5^-$. Both were determined to have nearly identical metal–alkyl carbon bond distances (within esd's) (24)(Darensbourg D. J.; Bauch, C. G.; Rheingold, A. L. Inorg. Chem., in press). However, only the anion has been found to undergo CO_2 insertion.

The mechanism of the CO_2 reaction is believed to proceed through an associative interchange (I_a) type mechanism with a transition state similar to that shown in Figure 2. The presence of CO did not

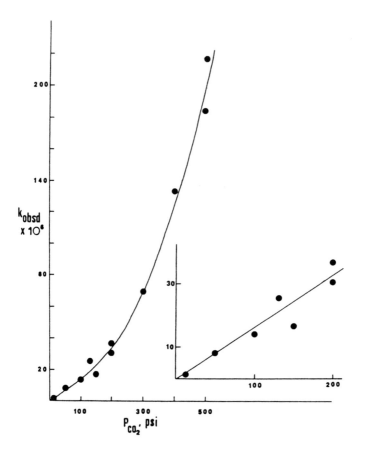

Figure 1. Plot of the pseudo-first-order rate constant (k_{obsd}) as a function of carbon dioxide pressure for the carboxylation of $CH_3W(CO)_5^-$ in THF at 25°C.

Figure 2. Transition state for CO_2 insertion into W-R bond.

Table II. Tungsten-CH_3 Bond Distances.

	d_{M-CH_3}
$(CO)_5W-CH_3^-$	2.325Å
cis-$(Me_3P)(CO)_4W-CH_3^-$	2.18Å
$(CO)_5Re-CH_3$	2.308Å

retard the rate of CO_2 insertion (**vide supra**) indicating that prior loss of coordinated CO was not involved. The activation parameters for CO_2 insertion into $CH_3W(CO)_4P(OMe)_3^-$ were determined ([20]). The ΔH^* and ΔS^* of 42.68 kj mole^{-1} and -181.2 eu respectively were indicative of a mechanism involving a highly ordered transition state with little net bond breaking.

A backside attack of CO_2, on the alkyl carbon, analogous to the SO_2 insertion mechanism was ruled out by studies of the α—carbon stereochemistry upon insertion ([25]). The insertion of CO_2 into the metal carbon bond of **threo-cis**-$W(CO)_4(L)(CHD-CHD-Ph)^-$ (L = CO and PMe_3) proceeds with retention of configuration at the α-carbon (Scheme 1) ([26]). This is in contrast to the inversion of configuration at the alpha carbon found in backside SO_2 insertion reactions.

Although CO doesn't retard the rate of CO_2 insertion it does compete with CO_2 for a concurrent insertion reaction into the metal-carbon bond ([23]). Thus, the anionic alkyl complexes of the group 6 metals provide a unique opportunity to compare carbonylation and carboxylation reaction under comparable conditions. The reactivity of $RW(CO)_5^-$ toward CO parallels that of $RMn(CO)_5$ ([27–31]). The mechanistic aspects of the carbonylation and carboxylation reaction are summarized and compared in Table III. In studies involving simultaneous insertions of both CO and CO_2 with $CH_3W(CO)_5^-$, it was found that the rate of disappearance of the alkyl complex was approximately equal to the sum of the individually determined rate parameters. Hence, the two processes occur simultaneously and independently of one another. These dissimilarities (dependences on metal, R group, and ancillary ligands) in CO_2 and CO insertion processes can be exploited in preferentially affecting carbon-carbon bond forming reactions involving carbon dioxide in the presence of carbon monoxide. The consequence of the mechanistic differences should be of major importance in catalytic processes designed to utilize carbon dioxide as a source of chemical carbon.

Metal Alkoxides. Notwithstanding our understanding of the mechanistic aspects operative in insertion reaction of carbon dioxide with metal-hydride and metal-carbon (alkyls or aryls) bonds has

Table III. Summary of Mechanistic Aspects of Carbonylation vs
Carboxylation Reactions.

Reaction Variables	Carboxylation	Carbonylation[a]
Kinetic order in CO_2 or CO	First-order in CO_2	Mixed-order in CO; independent of CO at high CO pressures
Nature of Metal	Third row more reactive than first row	First row more reactive than third row
R dependence	Small dependence on R group, alkyls faster than aryls	Reaction greatly retarded by electron-withdrawing R substituents
Ancillary ligands	Sterically nonencumbering phosphorus donor ligands greatly accelerate reaction	Little effect
Stereochemistry at α-carbon	Retention of configuration	Retention of configuration

[a] These observations have been extensively noted for $RMn(CO)_5$, and the more limited study on the group 6 anionic analogs reported herein is in complete agreement with these generalizations.

undergone significant advancement, analogous processes involving
metal-alkoxides have been less well-interpreted (32–34). We have
recently synthesized $W(CO)_5OR^-$ (R = Ph, $C_6H_4CH_3-m$) derivatives from
$W(CO)_5THF$ and $[Et_4N][OR]$ salts (Darensbourg, D. J.; Sanchez, K. M.;
Rheingold, A. L. J. Am. Chem. Soc., in press). In solution these
species are extremely CO labile, decomposing readily to
$W_4(CO)_{12}(\mu_3-OR)_4^-$ in the absence of a CO atmosphere (35).
Treatment of $[Et_4N][W(CO)_5OR]$ with CO_2 in THF under mild
conditions (< 1 atmosphere CO_2 at ambient temperature) results in
rapid production of $W(CO)_5OC(O)OR^-$, the product of CO_2 insertion into
the W-OR bond. The tungsten pentacarbonyl aryl carbonate complex was
characterized by IR, ^1H, and ^{13}C NMR spectroscopies. Addition of
small quantities of water to solutions of $W(CO)_5OC(O)OR^-$ causes
bright orange crystals of $[Et_4N]_2[W(CO)_4(\eta^2-CO_3)]\cdot H_2O$ to precipitate
from solution. The molecular structure of the tungsten tetracarbonyl
carbonate was unambiguously assigned on the basis of an X-ray
crystallographic determination. This chemistry is summarized in
Scheme 2 (Darensbourg, D. J.; Sanchez, K. M.; Rheingold, A. L. J. Am.
Chem. Soc., in press).
Because of the extreme CO lability of the $W(CO)_5OR^-$ species, CO
loss might be a prerequisite for CO_2 insertion. However, the rate of
CO_2 insertion is not inhibited by the presence of carbon monoxide.
Hence, we believe that an open coordination site is unnecessary for
the insertion process to occur. The reaction is thought to involve a
concerted insertion process, similar to that proposed and well
documented for the insertion of CO_2 into $CH_3W(CO)_5^-$. Nevertheless,
insertion of CO_2 into the W-OR bond is more facile than the
corresponding process involving W-R.
Attempts to induce the insertion of CO into the W-OPh bond of
$(CO)_5WOPh^-$ have failed at CO pressures up to 500 psi. In situ high
pressure FTIR measurements revealed only the presence of the starting
phenoxide tungsten complex.

Catalytic Processes Using CO_2 Promoted by Transition Metal Complexes

In this section we wish to incorporate our knowledge of the scope of
transition metal-CO_2 chemistry in exploring potential uses of CO_2 as
a feedstock in the synthesis of fine and bulk chemicals.

Alkyl Formate Production. In the past few years, formate esters have
become an important class of organic compounds mainly because of
their versatility as chemical feedstock (16,36–42), and as raw
materials for the perfume and fragrance industry (43–46).
Specifically, formate esters (methyl, ethyl, pentyl, etc.) have been
used as starting material for the production of aldehydes (36),
ketones (36), carboxylic acids (37–40), and amides (42). For
example, methyl formate can be hydrolyzed to formic acid (39,40) or
catalytically isomerized to acetic acid (38). On the other hand,
alkyl formates have been employed in the perfume and fragrance
industry in amounts of approximately 1000 to 3000 lb/year (43–46).
Among the formates that have been commonly used for these purposes
are: octyl (43), heptyl (44), ethyl (45), and amyl (46) formates.
Our recent interest in the chemistry of carbon dioxide (9,12)
has included investigations of the synthesis of alkyl formates
utilizing CO_2 as a source of chemical carbon (Equation 7) (47,48).

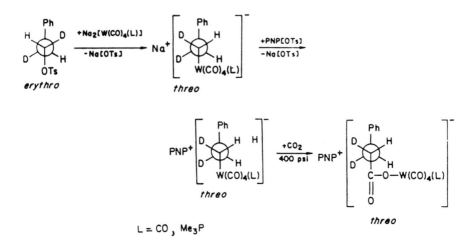

$$L = CO, \, Me_3P$$

Scheme 1.

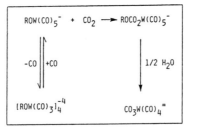

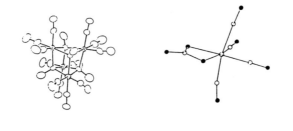

$$[ROW(CO)_3]_4^{-4} \qquad\qquad CO_3W(CO)_4^{=}$$

Scheme 2.

$$CO_2 + H_2 + ROH \xrightarrow{[cat]} HCO_2R + H_2O \qquad (7)$$

The catalysts or catalysts precursors employed in these studies were anionic group 6 carbonyl complexes (50) or group 8 metal carbonyl clusters (37–40), where reaction conditions were 500 psi (CO_2/H_2) and 125° C. For the group 6 metal catalysts, the turn-over numbers obtained for the methyl formate production were ca. 15 using methanol as solvent for a 24 hour period. The anionic metal carbonyls examined as catalysts precursors included: $HM_2(CO)_{10}^-$, $HCO_2M(CO)_5^-$, and $CH_3CO_2M(CO)_5^-$ as their PPN salts (PPN = bis(triphenylphosphine)-iminium and M = Cr or W). The proposed reaction pathway is depicted in Scheme 3.

Congruous with the catalytic cycle represented in Scheme 3, the metalloformate derivatives are extremely CO labile, specifically at equatorial CO sites as demonstrated by ^{13}CO labelling studies (equation 8). The dissociation of CO is important in catalysis, for the additon of CO inhibits the catalytic hydrogenation of carbon dioxide. The nature of the species (boxed) in Scheme 3 is thought to involve no change in the metal's oxidation state, i.e., a ligand-assisted heterolytic splitting of dihydrogen (49–51). That the distal oxygen atom of monodentate formates or carboxylates is extremely basic, is seen in its ability to interact with kryptofix–221 encapsulated sodium ion (Figure 3) (52).

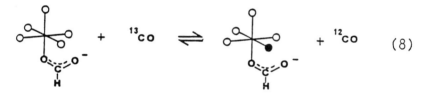

$$(8)$$

Carboxylation of methanol (Equation 9), where carbon monoxide orginates from the reverse water-gas shift reaction (Equation 4), was ruled out as a possible route to methyl formate in this instance. This conclusion is based on the observation that when $(^{13}CO)_5WO_2CH^-$ was utilized as the catalyst in the presence of $^{12}CO_2$, only $H^{12}CO_2Me$ was detected initially **via** GC-MS. Additionally, when $CH_3CO_2W(CO)_5^-$ was employed as the catalyst the first formed product in quantitative yield was CH_3CO_2H, with subsequent esterification affording $CH_3CO_2CH_3$.

$$CO + MeOH \longrightarrow HCO_2Me \qquad (9)$$

Although we have been able to demonstrate that methyl formate is derived **directly** from carbon dioxide, it is possible, employing the same metal carbonyl catalyst precursors, to catalyze the production of methyl formate from the reaction of CO and methanol (Equation 9). A catalytic cycle consistent with our current knowledge in this area is represented in Scheme 4 (53)(Darensbourg, D. J.; Gray, R. L.; Ovalles, C. J. Molec. Catal., in press). It is relevant to note herein that carbon dioxide is a poison to the catalytic process diagrammed in Scheme 4 in that it reacts with the co-catalyst, methoxide, to produce methyl carbonate.

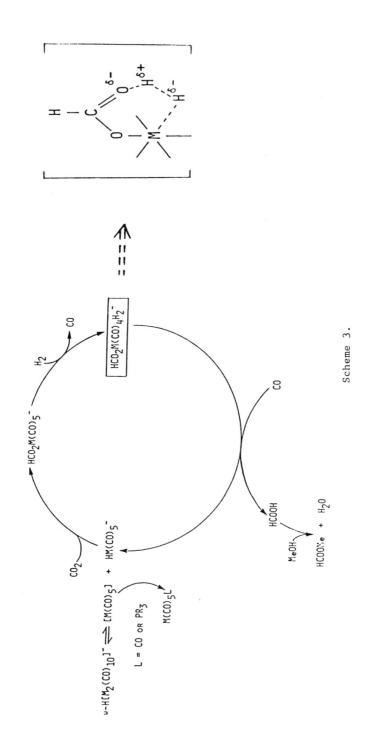

Scheme 3.

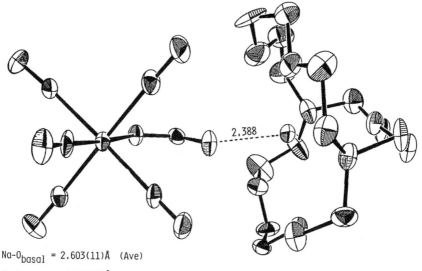

Na-O$_{basal}$ = 2.603(11)Å (Ave)

Na-O$_{apical}$ = 2.542(8)Å

Figure 3.   ORTEP drawing of [Na-kryptofix-221][W(CO)$_5$O$_2$CH].

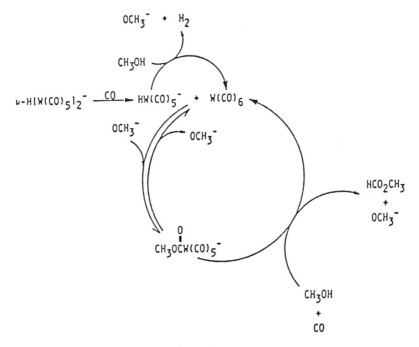

Scheme 4.

Mechanistic aspects of carbon dioxide hydrogenation processes (Equation 7) carried out in solvents other than methanol, e.g., ethanol or propanol, were completely consistent with those noted in methanol solvent. However, the catalytic activities were considerably lower in these alcohols as compared with methanol, by a factor of ~3. This effect was attributed to a solvent inhibition of the addition of dihydrogen to the unsaturated metal species. For this reason, we have turned our attention towards the reaction of alkyl halides with CO_2 and H_2 in order to provide a more effective pathway to higher molecular weight alkyl formates (butyl or octyl) using CO_2 as a source of carbon (Equation 10). Anionic group 6 metal complexes were used as catalysts, and the presence of a sodium salt ($NaHCO_3$ or $NaOCH_3$) was required in order to regenerate the catalytically active intermediates (Darensbourg, D. J.; Ovalles, C. J. Am. Chem. Soc., in press). A general catalytic cycle for the production of formate esters starting from alkyl halides, CO_2, and H_2 is shown in Scheme 5.

$$RX + CO_2 + H_2 \xrightarrow[\text{NaY}]{\text{[cat]}} HCO_2R + NaX + HY \qquad (10)$$

Mechanistic aspects of this catalytic process have been forthcoming from kinetic investigations of the component reactions which comprise the proposed cycle. The rate-limiting process in the catalytic cycle is the reaction between the anionic halide complexes,

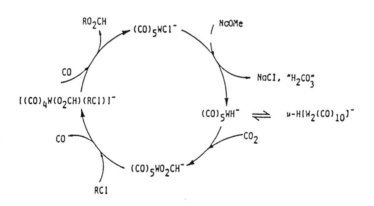

Scheme 5.

$M(CO)_5X^-$, and H_2 in the presence of a general base to provide anionic metal hydrides. This process was shown to be first-order in both metal complex and dihydrogen and was not inhibited by addition of carbon monoxide. Consistent with the rds in catalysis being formation of the metal hydride intermediate, the metal catalyzed reaction of $RX/CO_2/H_2$ to provide HCOOR is not inhibited by CO. The well-established formation of metalloformate, $M(CO)_5O_2CH^-$, from $M(CO)_5H^-$ and CO_2 is followed by a less facile process involving the reaction of the metalloformate with RX. This latter reaction is first-order in both metal complex and alkyl halide and is inhibited by carbon monoxide.

CO_2 Methanation. Finally, we have studied the catalytic activity and selectivity toward methanation of carbon dioxide using several alumina supported ruthenium clusters including $Ru_3(CO)_{12}$, $KHRu_3(CO)_{11}$, $[PPN][HCO_2Ru_3(CO)_{10}]$, $H_4Ru_4(CO)_{12}$, $KH_3Ru_4(CO)_{12}$, $[PPN][H_3Ru_4(CO)_{12}]$, and $Ru_6C(CO)_{17}$ (54). Comparative studies were made with the mononuclear complexes $RuCl_3$ and $Ru(CO)_5$. The latter species provides a low-valent, organometallic, mononuclear ruthenium source. Catalysts were supported by impregnation over alumina (partially dehydroxylated at 150° C **in vacuo**) and activated in hydrogen at 200° C. Catalyst characterization included diffuse reflectance infrared spectroscopy, surface area determination, and electron microscopy. In general, the cluster derived catalysts were more active than the analogously prepared catalyst obtained from $RuCl_3$, e.g., at 180° C the catalyst derived from $Ru_6C(CO)_{17}$ was 22 times more active than that derived from $RuCl_3$. The activity of the supported neutral species was observed to increase as the number of ruthenium atoms present in the precursor complex increase, i.e., $Ru(CO)_5 < Ru_3(CO)_{12} < Ru_6C(CO)_{17}$.

Acknowledgments. The authors are most grateful to the National Science Foundation, whose support has made possible their contributions to the research described herein. They are likewise appreciative to all their colleagues mentioned in the references, whose many original contributions have made this such an exciting area of research to be in.

Literature Cited

1. Henrici-Olive, G.; Olive, S. J. Mol. Cat. 1978, 3, 443.
2. Henrici-Olive, G.; Olive, S. Angew. Chem. Int. Ed. 1976, 15, 136.
3. Masters, C. Adv. Organomet. Chem. 1979, 17, 61.
4. Henrici-Olive, G.; Olive, S. The Chemistry of the Catalyzed Hydrogenation of Carbon Monoxide, Springer-Verlag; Berlin, Heidelberg, Germany,1984.
5. Keim, W. Catalysis in C_1 Chemistry; Keim W., Ed.; Reidel D. Publishing Co., Dorecht, Holland, 1983, p. 5.
6. Eisenberg, R.; Hendricksen, D. E. Adv. Catal. 1979, 28, 79.
7. Denise, B.; Sneeden, R. P. Chem. Tech. 1982, 12, 108.
8. Lapidus, A. L.; Ping, Y. Y. Russ. Chem. Rev. 1981, 50, 63.
9. Darensbourg, D. J.; Kudaroski, R. A. Adv. Organomet. Chem. 1983, 22, 129.

10. Sneeden, R. P. A. In Comprehensive Organometallic Chemistry; Wilkinson, G.; Stone, F. G. A.; Abel, E. W., Eds.; Pergamon Press: Oxford, 1982; Vol. 8, p.225.

11. Ito, T.; Yamamoto, A. Organic and Bio-Organic Chemistry of Carbon Dioxide; Inoue, S.; Yamazaki, N., Eds.; Kondonsha, Ltd.: Toyko, Japan, 1982, p 79.

12. Darensbourg, D. J.; Ovalles, C. Chem. Tech. 1985, 15, 636.

13. Chem. Eng. News, 1985, 63, 10.

14. Czoukowski, M. P.; Bayne, A. R. Hydrocarbon Process. 1980, 59, 103.

15. Leonard, J. D. U.S. Patent 4 299 981 (1981).

16. Ikaraski, T. Chem. Economy & Eng. Rev. 1980, 12, 31

17. Darensbourg, D. J.; Rokicki, A. Organometallics 1982, 1, 1685.

18. Slater, S. G.; Lusk, R.; Schumann, B. F.; Darensbourg, M. Organometallics 1982, 1, 1662.

19. Darensbourg, D. J.; Rokicki, A.; Darensbourg, M. Y. J. Am. Chem. Soc. 1981, 103, 3223.

20. Kao, S. C.; Darensbourg, M. Y.; Schenck, W. Organometallics 1984, 3, 871.

21. Darensbourg, D. J.; Pala, M.; Waller, J. Organometallics 1985 2, 1285.

22. Darensbourg, D. J.; Kudaroski, R. J. Am. Chem. Soc. 1984, 106, 3672.

23. Darensbourg, D. J.; Hanckel, R. K.; Bauch, C. G.; Pala, M.; Simmons, D.; White, J. N. J. Am. Chem. Soc. 1985, 107, 7463.

24. Rankin, D. W. H.; Robertson, A. Organomet. Chem. 1976, 105, 331.

25. Wojcicki, A. Acc. Chem. Res. 1971, 4, 344 and references therein.

26. Darensbourg, D. J.; Grötsch, G. J. Am. Chem. Soc. 1985, 107, 7473.

27. Wojcicki, A. Adv. Organomet. Chem. 1973, 11, 87.

28. Wojcicki, A. Adv. Organomet. Chem. 1974, 12, 31.

29. Calderazzo, F. Agnew. Chem. Int. Ed. Engl. 1977, 16, 299.

30. Kuhlman, E. J.; Alexander, J. J. Coord. Chem. Rev. 1979, 172, 405.

31. Flood, T. C. Top. Inorg. Organomet. Stereochem. 1981, 12, 37.

32. Newman, L. J.; Bergman, R. G. J. Am. Chem. Soc. 1985, 107, 5314.

33. Crutchely, R. J.; Powell, J.; Faggiani, R.; Lock, C. J. L. Inorg. Chim. Acta 1977, 24, L15.

34. Tsuda, T.; Sanada, S.; Ueda, K.; Saegusa, T. Inorg. Chem. 1985, 24, 3465, and references therein.

35. McNeese, T. J.; Mueller, T. E.; Wierda, D. A.; Darensbourg, D. J.; Delord, T. E. Inorg. Chem. 1985, 24, 3465, and references therein.

36. Trecker, D. J.; Sander, M. R. U. S. Patent 4093661, 1973.

37. Pruett, R. L. Eur Pat. Appt. EP 45637A1, 1982, CA# 180799f(96).

38. Pruett, R. L.; Kacmarick, R. T. Organometallics 1982, 59, 103.

39. Czarkaroski, M. P.; Bayne, A. R. Hydrocarbon Process. 1980, 59, 103.

40. Leonard, J. D.; U. S. Patent 4299981, 1981.

41. Dryvy, D. J.; Williams, P. S. Eur. Pat. Appt. EP106656A 1984; CA# 90427a (101).

42. Green, M. J. Eur. Pat. Appt. EP107441A1, 1984, CA# 101378r (101).

43. Opdyke, D. L. J. Food Cosmetl. Toxicol. 1979, 17 (suppl. 1), 883.
44. Ibid., 1978 16(suppl. 1), 771.
45. Ibid., 1978 16(suppl. 1), 737.
46. Ibid., 1980 18(6), 649.
47. Darensbourg, D. J.; Ovalles, C. J. Am. Chem. Soc. 1984, 106, 3750.
48. Darensbourg, D. J.; Ovalles, C.; Pala, M. J. Am. Chem. Soc. 1983, 105, 5937.
49. Halpern, J.; Milne, J. B. Second International Congress on Catalysis; Technip: Paris, 1961; p 445.
50. Fryzuk, M. D.; MacNeil, P. A. Organometallics 1983, 2, 682.
51. James, B. R. In Comprehensive Organometallic Chemistry; Wilkinson, G.; Stone, F. G. A.; Abel, E. W., Eds.; Pergamon Press Ltd: Great Britain, 1982; Vol. 8, p 285.
52. Darensbourg, D. J.; Pala, M. J. Am. Chem. Soc. 1985, 107, 5687.
53. Darensbourg, D. J.; Gray, R. L.; Ovalles, C.; Pala, M. J. Molec. Catal. 1985, 29, 285.
54. Darensbourg, D. J.; Ovalles, C. Inorg. Chem. 1986, 25, 1603.

RECEIVED April 21, 1987

Chapter 5

Use of Stoichiometric Reactions in the Design of Redox Catalyst for Carbon Dioxide Reduction

Daniel L. DuBois and Alex Miedaner

Solar Energy Research Institute, Golden, CO   80401

By using a combination of known reactions, synthetic chemistry, and electrochemistry, new catalysts for the electrochemical reduction of CO_2 have been developed. Consideration of the electrochemical properties of a variety of transition metal complexes containing polyphosphine ligands led to a more detailed study of metal complexes of the type $[M(PP_2)(PR_3)](BF_4)_2$ (where M is Ni, Pd, and Pt; PP_2 is $PhP(CH_2CH_2PPh_2)_2$; and R is C_2H_5 or OCH_3). When M is Pd, these complexes are active catalysts for the reduction of CO_2. Mechanistic details of the catalytic reactions are briefly discussed.

The combination of well known stoichiometric reactions into a catalytic cycle is a logical approach to the development of catalysts. Perhaps the best known example of this approach in homogeneous catalysis is the development of the Wacker process for the oxidation of ethylene to acetaldehyde (1-2). In this paper we would like to demonstrate the use of this approach in the development and synthesis of homogeneous catalysts for the electrochemical reduction of CO_2. Other homogeneous catalysts for the electrochemical reduction of CO_2 have been reported. These include porphyrins and tetraazamacrocyclic complexes (3-7), $Rh(dppe)_2Cl$ (4), iron sulfur clusters (9), bipyridine complexes (10-12), and formate dehydrogenase (13). The catalysts reported in this paper appear to have different mechanistic features than the homogeneous catalysts reported previously and operate at potentials approximately 0.6 V positive of other homogeneous catalysts with the exception of formate dehydrogenase.

Review of Some Known Chemical Reactions Related to CO_2 Reduction

Scheme 1 represents a possible catalytic cycle for the electrochemical reduction of CO_2 to formate. (In this Scheme and in Scheme 2, L represents an undefined ligand and M represents a

transition metal.) The catalytic cycle depicted in Scheme 1 is
based on a combination of known chemical reactions. For example,
the protonation of many low valent transition metal complexes leads
to the formation of metal hydrides as shown in Step 1 of Scheme 1
(14). The types of metal complexes which most readily protonate to
form metal hydrides are metal phosphine complexes and
organometallic complexes. Although this reaction has been used
primarily in the synthesis of metal hydrides, its potential use in
electrocatalytic cycles is illustrated in Scheme 1.

The reaction of transition metal hydrides and metal alkyls
with CO_2 frequently results in the formation of metal formates and
carboxylates via an insertion of CO_2 into a metal hydride or metal
carbon bond, Step 2 of Scheme 1 (15-19). In some instances, the
mechanism for this reaction has been investigated in detail. It
has been found that the reaction can proceed by either a
dissociative mechanism to produce a coordinatively unsaturated
metal hydride as an intermediate, or it can occur by an associative
mechanism (20-25). Thus, the metal hydride shown in Scheme 1 may
or may not be required to be coordinatively unsaturated.
Organometallic and metal phosphine complexes are again the two
classes of complexes most commonly involved in CO_2 insertions into
metal hydrogen bonds (15-19).

The acid cleavage of a metal oxygen bond in metal
carboxylates and formates is a commonly used method for the
preparation of formic and carboxylic acids. The facility of this
reaction depends on the strength of the metal oxygen bond. Since
it is desirable that this bond be cleaved rapidly in a catalytic
cycle, our attention was restricted to metal complexes of the Fe,
Co and Ni triads for which the metal oxygen bond is expected to be
weaker than those of the early transition metals.

Since the overall reaction is a two-electron reduction, it is
desirable for the oxidized metal complex to undergo a reversible
two-electron reduction as shown in Step 4 of Scheme 1. At pH 7 the
standard potential for the reduction of CO_2 to formic acid is -0.61
vs NHE (26) or -0.85 vs SCE. Since it would be desirable if the
reduction of CO_2 could ultimately be carried out in aqueous

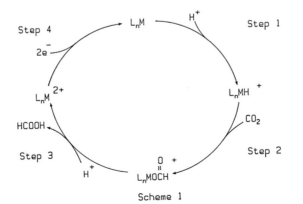

Scheme 1

solutions under neither strongly acidic nor strongly basic conditions, i.e., at pH values between 4 and 10, the potential for the reduction shown in Step 4 should ideally be between -0.5 and -1.1 V vs SCE.

Electrochemical Studies of Polyphosphine Metal Complexes

In many instances redox processes of transition metal complexes containing monodentate phosphine or carbonyl ligands are irreversible (27-34). Frequently when monodentate phosphine ligands are replaced with bidentate phosphine ligands, the redox processes become more reversible presumably because metal phosphorus bond cleavage and bond formation is repressed (35-43). This increased electrochemical reversibility should also be true for metal complexes containing polydentate phosphine ligands. For this reason, we carried out the syntheses of Fe, Co, and Ni complexes containing polydentate phosphine ligands and investigated their electrochemical behavior (44). Some of these complexes and their electrochemical parameters are shown in Table I. As can be seen from Table I, the Fe(II) complexes are reduced in two sequential one-electron steps with the (I/0) couple lying outside of the desired potential range. The large negative potentials suggest that the iron complexes would not be good candidates for CO_2 reduction at reasonable pH values. Similarly, the electrochemical parameters for the cobalt complexes are not ideal

Table I. Comparison of Cyclic Voltammetry Data for $[ML(CH_3CN)_n](BF_4)_2$ Complexes

Complex	$E_{1/2}$ Values for Indicated Changes in Oxidation States[a]			
	III/II	II/I	I/0	0/-I
$[Fe(PP_3)(CH_3CN)](BF_4)_2$	+1.25(80)[b]	-1.12c[*],[c] -0.27a	-1.33(60)	
$[Fe(dppe)_2(CH_3CN)_2](BF_4)_2$	+1.36(90)	-1.11c* -0.24a*	-1.39(140)	
$[Co(PP_3)(CH_3CN)_2](BF_4)_2$	+1.07c* +0.24a*	-0.14(70)		
$[Co(dppe)_2(CH_3CN)](BF_4)_2$	+1.60* +0.27a*	-0.31(60)	-1.16(70)	-1.63(60)
$[Ni(PP_3)(CH_3CN)](BF_4)_2$		-0.63(100)	-0.88c*	
$[Ni(dppe)_2](BF_4)_2$		-0.31(60)	-0.49(60)	

[a] All potentials are reported vs SCE and all measurements were carried out in acetonitrile solutions ~0.2N in NEt_4BF_4.
[b] Value in parenthesis indicates peak to peak separation.
[c] The letters a and c for irreversible redox waves (*) indicate whether the peak is anodic or cathodic, respectively. The potentials listed for irreversible couples represent the potential of the peak current and not $E_{1/2}$. See Legend of Symbols at end of paper for ligand abbreviations.

for CO_2 reduction. However, the half wave potentials for the nickel complexes for the Ni(II/I) and Ni(I/0) couples are in the range of interest. In addition, the small difference in the potentials for the Ni(II/I) and Ni(I/0) couples made further investigation of complexes of the nickel triad desirable. In Table II are shown a number of Ni, Pd and Pt complexes which we have synthesized, characterized, and studied electrochemically. From the electrochemical parameters given in this Table, it can be seen that, in general, the nickel complexes undergo two reversible one-electron reductions while the palladium and platinum complexes undergo reversible or quasi-reversible two-electron reductions. The number of electrons associated with each of the waves have been determined by controlled potential electrolysis experiments. The half wave potentials for these redox processes are generally within the desired potential range of -0.5 to -1.1 V.

Two classes of metal complexes are shown in Table II. One class is composed of metal complexes of the type $[M(\text{diphosphine})_2](BF_4)_2$ and the other of complexes of the type $[M(\text{triphosphine})(PR_3)](BF_4)_2$. These two classes of complexes are quite similar in their electrochemical properties as shown in Table II. However, as described below, their chemical properties are quite different in terms of their ability to catalyze the electrochemical reduction of CO_2.

The complexes of the type $[M(PP_2)(PR_3)](BF_4)_2$ (where M is Ni, Pd, and Pt; PP_2 is $PhP(CH_2CH_2PPh_2)_2$; and R is C_2H_5 or OCH_3) are all new complexes and have been characterized by ^{31}P and ^1H NMR spectroscopy, infrared spectroscopy, and elemental analysis in addition to the electrochemical studies described above. The ^{31}P NMR spectrum of $[Pd(PP_2)P(OMe)_3](BF_4)_2$ is shown in Figure 1 along with the simulated NMR spectrum. All of the spectral and analytical data are consistent with the formulation of this complex and the other complexes in Table II as square-planar metal complexes.

Table II. Cyclic Voltammetry Data for $[M(\text{diphos})_2](BF_4)_2$ and $[M(PP_2)(PR_3)](BF_4)_2$ Complexes

Compound	$E_{1/2}(II/I)^a$	$E_{1/2}(II/0)$	$E_{1/2}(I/0)$
$[Ni(dppe)_2](BF_4)_2$	-0.31(60)		-0.49(60)
$[Ni(PP_2)(PEt_3)](BF_4)_2$	-0.37(90)		-0.65(60)
$[Ni(PP_2)P(OMe)_3](BF_4)_2$		-0.45(55)	
$[Pd(dppe)_2](BF_4)_2$		-0.64(35)	
$[Pd(dppp)_2](BF_4)_2$		-0.45(35)	
$[Pd(PP_2)(PEt_3)](BF_4)_2$		-0.72(43)	
$[Pd(PP_2)P(OMe)_3](BF_4)_2$		-0.61(60)	
$[Pt(PP_2)(PEt_3)](BF_4)_2$		-0.93(45)	

^{a}All potentials are given in volts vs SCE and all measurements were carried out in acetonitrile solutions ~0.3N in NEt_4BF_4. The number in parentheses indicates the difference in potential for the peak current for the cathodic and anodic waves. See Legend of Symbols at end of paper for ligand abbreviations.

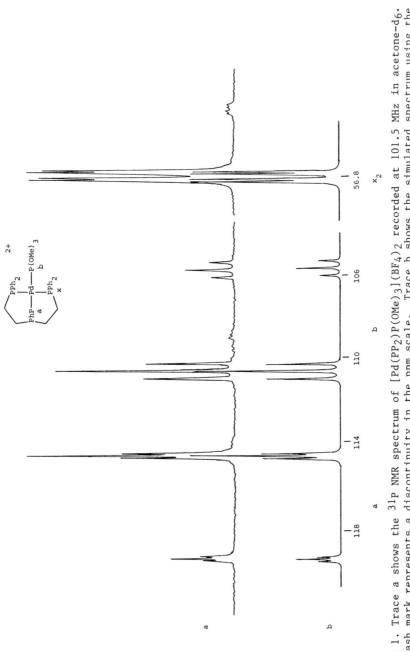

Figure 1. Trace a shows the ^{31}P NMR spectrum of $[Pd(PP_2)P(OMe)_3](BF_4)_2$ recorded at 101.5 MHz in acetone-d_6. The slash mark represents a discontinuity in the ppm scale. Trace b shows the simulated spectrum using the following parameters: P_a, 116.17 ppm; P_b, 108.81 ppm; P_x, 56.82 ppm; J_{ab} = 490 Hz; J_{ac} = 9 Hz, J_{bc} = 37 Hz.

Catalytic Studies

The complexes of Table II were evaluated for their catalytic activity by a series of cyclic voltammetry experiments such as those shown in Figure 2 for $[Pd(PP_2)(P(OMe)_3](BF_4)_2$. Trace a of Figure 2 shows the cyclic voltammogram of this complex under N_2 or CO_2. The observation of a fully reversible wave in the presence of CO_2 indicates that the reduced palladium complex, $[Pd(PP_2)P(OMe)_2]$, does not react with CO_2. This observation is supported by the fact that analytically pure $[Pd(PP_2)P(OMe)_3]$, synthesized by the reduction of $[Pd(PP_2)P(OMe)_3](BF_4)_2$ with hydrazine, does not react with CO_2 (45). When HBF_4 is added to acetonitrile solutions of $[Pd(PP_2)P(OMe)_3](BF_4)_2$, the reduction wave is irreversible as shown in Trace b of Figure 2. This is consistent with the palladium(0) complex formed on reduction reacting with HBF_4 to form a hydride complex. Addition of CO_2 to an acidic acetonitrile solution of $[Pd(PP_2)(P(OMe)_3](BF_4)_2$ results in a further increase in the peak current as shown in Trace c of Figure 2. This increase in current in the presence of CO_2 is attributed to catalytic CO_2 reduction. This interpretation is supported by the observation that up to 5 moles of CO per mole of $[Pd(PP_2)P(OMe)_3](BF_4)_2$ are detected by a GC analysis of the gas phase above the solution in bulk electrolysis experiments. Current efficiencies of up to 75% for the production of CO have been observed. No catalytic activity is observed for any of the Ni or Pt complexes. This observation illustrates the importance of the metal in this catalytic reaction. In addition, the $[Pd(diphosphine)_2](BF_4)_2$ complexes are not catalysts under these conditions. The failure of the latter complexes to catalyze the reduction of CO_2 is attributed to their inability to dissociate a phosphine ligand. This hypothesis is supported by the observation that addition of trimethylphosphite to a solution of $[Pd(PP_2)P(OMe)_3](BF_4)_2$ inhibits its ability to reduce CO_2.

Kinetic studies made on $[Pd(PP_2)(PEt_3)[(BF_4)_2$, and reported elsewhere (45), indicate that the rate of catalysis is first order in CO_2, first order in catalyst, and first order in acid at low acid concentrations. These results are consistent with the mechanism shown in Scheme 2. In comparison with Scheme 1, two important features should be noted. First in Scheme 2, the formation of a coordinatively unsaturated metal hydride complex is necessary for CO_2 insertion to occur. A priori there is no way of knowing whether or not the generation of a coordinatively unsaturated metal hydride will be required for catalysis since evidence exists for both associative and dissociative pathways for CO_2 insertion into metal hydride and metal carbon bonds (20-25). This is the reason that complexes of the types $[Pd(diphosphine)_2](BF_4)_2$ and $[Pd(PP_2)(PR_3)](BF_4)_2$ were prepared and evaluated. The presence of a monodentate phosphine ligand in the $[Pd(PP_2)(PR_3)](BF_4)_2$ complexes permits a phosphine ligand to dissociate during the catalytic cycle. At the same time, the use of a tridentate phosphine ligand allows electrochemical reversibility to be preserved.

A second difference between Scheme 1 and Scheme 2 is that the product formed in the catalytic cycle of Scheme 1 is formate, and in Scheme 2 the product is CO. This observation is interpreted in

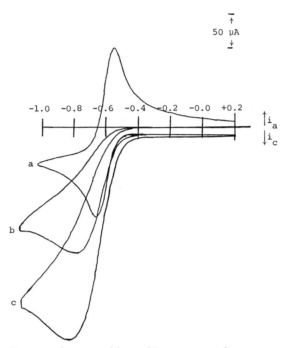

Figure 2. Trace a is a cyclic voltammogram of an approximately
1×10^{-3} M solution of $[Pd(PP_2)P(OMe)_3](BF_4)_2$ under an
atmosphere of either nitrogen or carbon dioxide. Trace b is a
cyclic voltammogram of the same solution after addition of HBF_4
to produce an approximately 1×10^{-2} M solution. Trace c
illustrates the increase in current which is observed when
carbon dioxide is bubbled through the acidic solution for five
minutes. The solutions were all 0.2 N NEt_4BF_4 in
acetonitrile. The working electrode was glassy carbon, the
counter electrode was a Pt grid, and the reference electrode
was SCE.

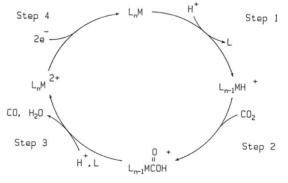

Scheme 2

terms of an insertion of CO_2 into the palladium hydride bond with formation of an intermediate metallocarboxylic acid as shown in Step 2 of Scheme 2. Subsequent protonation of this acid followed by loss of water would result in the formation of CO. The possibility of such an intermediate is supported by the existence of a number of complexes of this type (46-52). While Step 2 of Scheme 2 appears to be a major deviation from that proposed in Scheme 1, it probably is a reflection of a rather delicate balance between metal oxygen and metal carbon bond strengths.

The parallels between Schemes 1 and 2 illustrates that a knowledge of stoichiometric reactions can be utilized in the design of electrocatalysts. In particular, the use of electrochemical reductions to generate metal hydride complexes could result in a number of different types of catalytic reductions depending on the nature of the substrate.

Acknowledgment

This research was supported by the U.S. Department of Energy, Office of Basic Energy Sciences, Division of Chemical Sciences.

Legend of Symbols

dppe is $Ph_2PCH_2CH_2PPh_2$
dppp is $Ph_2PCH_2CH_2CH_2PPh_2$
PP_2 is $PhP(CH_2CH_2PPh_2)_2$
PP_3 is $P(CH_2CH_2PPh_2)_3$
M represents a transition metal.
L represents a monodentate ligand.

Literature Cited

1. Parshall, G. W. *Homogeneous Catalysis*; Wiley: New York, 1980; pp 102-104.
2. Cotton, F. A.; Wilkinson, G. *Advanced Inorganic Chemistry*; Wiley: New York, 1980; pp 1300-1302.
3. Eisenberg, R.; Fisher, B. *J. Amer. Chem. Soc.* 1980, 102, 7363.
4. Meshitsuka, S.; Ickikawa, M.; Tamaru, K. *J. Chem. Soc., Chem. Commun.* 1974, 158.
5. Takashi, K.; Hiratsuka, K.; Sasaki, H. *Chem. Lett.* 1979, 305.
6. Bradley, M. G.; Tysak, T.; Graves, D. J.; Vlachopoulos. *J. Chem. Soc., Chem. Commun.* 1983, 349.
7. Becker, J. Y.; Vainas, B.; Eger, R.; Kaufman, L. *J. Chem. Soc., Chem. Commun.* 1985, 1471.
8. Slater, S.; Wagenknecht, J. H. *J. Amer. Chem. Soc.* 1984, 106, 5367.
9. Tezuka, M.; Tajima, T.; Tsuchiya, A.; Matsumoto, Y.; Uchida, Y.; Hidai, J. *J. Amer. Chem. Soc.* 1982, 104, 6834.
10. Hawecker, J.; Lehn, M. M.; Ziessel, R. *J. Chem. Soc., Chem. Commun.* 1984, 328.
11. Ishida, J.; Tanaka, K.; Tanaka, T. *Chem. Lett.* 1985, 405.
12. Bolinger, C. M.; Sullivan, B. P.; Conrad, D.; Gilbert, J. A.; Story, N.; Meyer, T. J. *J. Chem. Soc., Chem. Commun.* 1985, 796.

13. Parkinson, B.; Weaver, P. F. Nature 1984, 309, 148.
14. Schunn, R. A. In Transition Metal Hydrides; Muetterties, E.
    L., Ed.; Dekker: New York, 1971; pp 218-221.
15. Darensbourg, D. J.; Kudaroski, R. A. Adv. Organomet. Chem.
    1983, 22, 129.
16. Denise, B.; Sneeden, R. P. A. CHEMTECH 1982, 12, 108.
17. Palmer, D. A.; Eldik, R. V. Chem. Rev. 1983, 83, 651.
18. Eisenberg, R.; Hendricksen, D. E. Adv. Catal. 1979, 28, 79.
19. Ito, T.; Yamamoto, A. In Organic and Bio-Organic Chemistry of
    Carbon Dioxide; Inoue, S.; Yamazaki, N., Eds.; Wiley: New
    York, 1980.
20. Darensbourg, D. J.; Rokicki, A.; Darensbourg, M. Y. J. Amer.
    Chem. Soc. 1981, 103, 3223.
21. Darensbourg, D. J.; Rokicki, A. J. Amer. Chem. Soc. 1982, 104,
    349.
22. Darensbourg, D. J.; Fisher, M. B.; Schmidt, R. E., Jr.;
    Baldwin, B. J. J. Amer. Chem. Soc. 1981, 103, 1297.
23. Darensbourg, D. J.; Kudaroski, R. J. Amer. Chem. Soc. 1984,
    106, 3672.
24. Darnesbourg, D. J.; Hanckel, R. K.; Bauch, C. G.; Pala, M.;
    Simmons, D.; White, J. N. J. Amer. Chem. Soc. 1985, 107, 7463.
25. Darensbourg, D. J.; Grotsch, G. Ibid. 1985, 107, 7423.
26. Encyclopedia of Electrochemistry of the Elements, Bard, A. J.,
    Ed.; Dekker: New York, 1976; Vol. 7, p 172.
27. Bontempelli, G.; Magno, F.; Schiavon, G.; Corain, B. Inorg.
    Chem. 1981, 20, 2579.
28. Bontempelli, G.; Magno, F.; Corain, B.; Schiavon, G. J.
    Electroanal. Chem. and Interfac. Electrochem. 1979, 103, 243.
29. Olson, D. C.; Keim, W. Inorg. Chem. 1969, 8, 2028.
30. Pilloni, G.; Valcher, S. J. Electroanal. Chem. and Interfac.
    Electrochem. 1972, 40, 63.
31. Pilloni, G.; Zotti, G.; Martelli, M. J. Electroanal. Chem. and
    Interfac. Electrochem. 1975, 63, 424.
32. Zotti, G.; Zecchin, S.; Pilloni, G. J. Organomet. Chem. 1983,
    246, 61.
33. Corain, B.; Bontempelli, G.; DeNardo, L.; Mazzocchin, G.-A.
    Inorg. Chim. Acta 1978, 26, 37.
34. Jasinski, R. J. Electrochem. Soc. 1983, 130, 834.
35. Sullivan, B. P.; Salmon, D. J.; Meyer, T. J. Inorg. Chem.
    1978, 17, 3334.
36. Pilloni, G.; Vecchi, E.; Martelli, M. J. Electroanal. Chem.
    and Interfac. Electrochem. 1983, 45, 483.
37. Pilloni, G.; Zotti, G.; Martelli, M. Inorg. Chem. 1982, 21,
    1284.
38. Kunin, A. J.; Nanni, E. J.; Eisenberg, R. Inorg. Chem. 1985,
    24, 1852.
39. Pilloni, G.; Zotti, G.; Martelli, M. J. Electroanal. Chem. and
    Interfac. Electrochem. 1974, 50, 295.
40. Zotti, G.; Zecchini, S.; Pilloni, G. J. Organomet. Chem. 1979,
    181, 375.
41. Martelli, M.; Pilloni, G.; Zotti, G.; Daolio, S. Inorg. Chim.
    Acta, 1974, 11, 155.
42. Bowmaker, G. A.; Boyd, P. D. W.; Campbell, G. K.; Hope, J. M.
    Inorg. Chem. 1982, 21, 1152.

43. Zotti, G.; Pilloni, G.; Rigo, P.; Martelli, M. J. Electroanal. Chem. and Interfac. Electrochem. 1981, 124, 277.
44. DuBois, D. L.; Miedaner, A. Inorg. Chem. 1986, 25, 000.
45. DuBois, D. L.; Miedaner, A. J. Amer. Chem. Soc. 1987, 109, 000.
46. Darensbourg, D. J.; Darensbourg, M. Y.; Walker, N.; Froelich, J. A.; Barros, H. L. C. Inorg. Chem. 1979, 18, 1401.
47. Grice, N.; Kao, S. C.; Pettit, R. J. Amer. Chem. Soc. 1979, 101, 1628.
48. Ford, P. C. Accts. Chem. Res. 1981, 14, 31.
49. Cheng, C. J.; Hendricksen, D. E.; Eisenberg, R. J. Amer. Chem. Soc. 1977, 99, 2791.
50. Deeming, A. J.; Shaw, B. L. J. Chem. Soc. A 1969, 443.
51. Appleton, T. G.; Bennett, M. A. J. Organomet. Chem. 1973, 55, C88.
52. Gibson, D. H.; Ong, T. S. Organometallics 1984, 3, 1911.

RECEIVED February 13, 1987

Chapter 6

Electrocatalytic Carbon Dioxide Reduction

B. Patrick Sullivan, Mitchell R. M. Bruce, Terrence R. O'Toole,
C. Mark Bolinger, Elise Megehee, Holden Thorp, and Thomas J. Meyer

University of North Carolina at Chapel Hill, Chapel Hill, NC 27514

This paper reviews recent work on the development of electrocatalysts for CO_2 reduction. Comparison of our electrocatalysts based on polypyridine complexes of the second and third row transition metals is made with previous work, and both areas are set in the framework of the known chemistry and electrochemistry of both uncoordinated CO_2 and CO_2-transitior metal complexes.

The emphasis of our work has been on mechanistic questions. For example, the family of complexes $\underline{fac}$-$[Re^I(bpy)(CO)_3L]^{n+}$ (where bpy is 2,2'-bipyridine and L is Br^-, Cl^- or CH_3CN) are facile stoichiometric or catalytic reagents that reduce CO_2 to CO, formate, or oxalate depending on the external conditions. Synthesis, electrochemical, and kinetic studies implicate the involvement of a minimum of five different pathways for this unusual system. A newly discovered electrocatalyt is the reactive metal hydride, $[Os(bpy)_2(CO)H]^-$, that has been found to reduce CO_2 by an associative mechanism yielding either CO or formate from a common intermediate. Related kinetic studies of fundamental steps in CO_2 activation or reduction have been conducted and their relationship to electrocatalytic CO_2 reduction has been

0097–6156/88/0363–0052$10.75/0

highlighted. Examples include CO_2 insertion into a metal-alkoxide or metal-hydride bond.

Finally, chemically modified electrodes have been prepared which allow the transposition of solution electrocatalytic chemistry to electrode surfaces. Although these studies are in their infancy it appears that new products (e.g., oxalate), and therefore new mechanistic pathways, have been found for some of the surface immoblized electrocatalsyts.

Photosynthetic reduction of carbon dioxide is a facile natural process even though the chemistry is complex and multielectron steps are required. Recent work has demonstrated that metal complexes play a crucial role (1) and it is anticipated that study of the homogeneous solution chemistry of CO_2 and its metal complexes will offer valuable clues to the mechanistic steps involved in biogenic CO_2 reduction. In fact, as in many problems in chemical reactivity, a detailed understanding of mechanism is key to the design of new catalytic systems. A mechanistic emphasis should also suggest advances in other areas such as routes to new energy sources, synthetic schemes for industrial chemicals, and methods for the removal of CO_2 as an atmospheric contaminant.

Even though detailed CO_2 reduction mechanisms are uncertain, and despite the fact that transition metal-CO_2 chemistry has developed slowly since the discovery of the first complexes, catalysts for abiogenic CO_2 reduction, especially electrocatalysts, have been found (2,3). Our own work in this area has led to the discovery of a series of electrocatalysts active for the reduction of CO_2 to CO, or formate (3a,b,c,s,w). More importantly, a fundamental grasp of chemical and electrochemical pathways and intermediates for a few select cases has been achieved, and in one instance, that of CO_2 insertion into a metal hydride complex, a detailed mechanistic picture has emerged.

These studies on CO_2 reduction are part of our larger effort on the activation and redox chemistry of small molecules and ions such as NO_3^{2-}, NH_3, O_2 and H_2O (4). Some long range goals of the work are to develop an understanding of the essential synthetic and mechanistic chemistry leading to reduction of carbon dioxide to formate, formaldehyde, methanol and oxalate and to use the results to develop new transition metal

electrocatalysts which will exhibit high product
selectivity, operate closer to the thermodynamic
potential of the appropriate redox process, and to
achieve the catalyzed reduction under conditions of high
current density.

Following a short review of the thermodynamics of
CO_2 reduction, CO_2 reduction at metal or carbon
electrodes, electrochemical and photoelectrochemical
reduction at semiconductor electrodes, CO_2-metal complex
reactivity, and the properties of related homogeneous
solution electrocatalytic systems, we will describe the
current scope of our attempts to develop new CO_2
electrocatalysts and describe the results of our
mechanistic studies.

THERMODYNAMICS OF CARBON DIOXIDE REDUCTION. A useful
summary of the thermodynamics of CO_2 reduction to
one-carbon fragments in aqueous solution under basic and
acidic conditions is shown in the Latimer-type diagram in
Scheme 1 (5). When referring to the diagram below,
recall that a negative potential means that the reduced
form of the couple is a better <u>reducing</u> agent than H_2,
and conversely, a positive value indicates that the
oxidized form is a better <u>oxidizing</u> agent than the proton
at the specified pH.

$$CO_2 \xleftrightarrow{-0.20} CO_2H \xleftrightarrow{-0.01} H_2C=O \xleftrightarrow{0.19} CH_3OH \xleftrightarrow{0.58} CH_4$$

$$\underset{\quad CO}{\vert\!\_\_\_\_} {}^{-0.1}\rceil \qquad\qquad [1M\ H^+]$$

$$CO_3{}^{2-} \xleftrightarrow{-1.0} HCO_2{}^- \xleftrightarrow{-.07} H_2C=O \xleftrightarrow{-0.6} CH_3OH \xleftrightarrow{-0.25} CH_4$$

$$[1M\ OH^-]$$

<u>Scheme 1</u>. (Values in Volts)

In acidic solution reduction of CO_2 to either formic
acid or CO is slightly endergonic with respect to the
H_2/H^+ redox couple, while reduction to methane is
actually spontaneous. Even though the various reductions
are accessible at reasonable potentials their kinetic
barriers can be quite severe, consequently, substantial
overpotentials can be incurred at the electrode surface.

ELECTROCHEMISTRY OF CO_2 REDUCTION. PRODUCTS AND
OVERPOTENTIAL REQUIREMENTS. A number of electrochemical
studies at metal or carbon electrodes have documented the
large overvoltages required for CO_2 reduction, both for
aqueous and non-aqueous media (6). Typical reduction
potentials required at either Pt or Hg working electrodes
are; -2.21V for dimethylformamide, -2.16V for H_2O at pH
7, and -2.2 to -2.7V for CH_3CN (using alkylammonium salts
as supporting electrolytes with the NaCl saturated sodium
chloride electrode as reference; SSCE, i.e., +0.24V
versus NHE) (6a). Under these conditions CO_2^- is the
initial reduction product, and once formed, is
exceedingly reactive.

The fate of electrochemically generated CO_2^- in
water depends upon the pH and the electrode composition.
Scheme 2 shows the potentials that have been used in the
preparative electroreduction of CO_2 at Pb, Pt, or Hg
electrodes, along with the ultimate products
(6a,b,h,j,l,o,s).

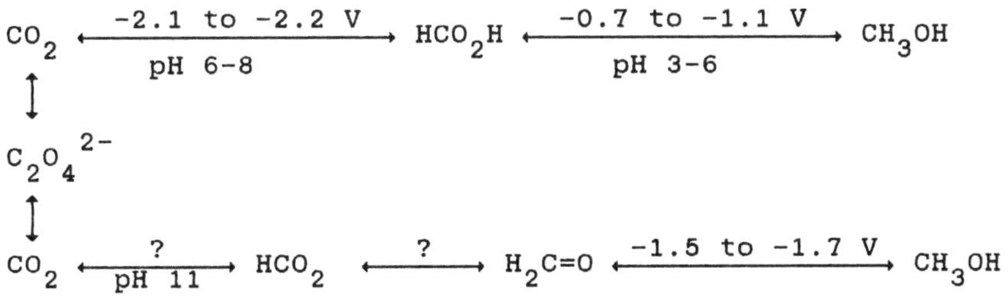

$$CO_2 \xleftarrow[\text{pH 6-8}]{\text{-2.1 to -2.2 V}} HCO_2H \xleftarrow[\text{pH 3-6}]{\text{-0.7 to -1.1 V}} CH_3OH$$

$$\updownarrow$$

$$C_2O_4^{2-}$$

$$\downarrow$$

$$CO_2 \xleftarrow[\text{pH 11}]{?} HCO_2 \xleftarrow{?} H_2C=O \xleftarrow{\text{-1.5 to -1.7 V}} CH_3OH$$

Scheme 2.

Of particular interest is that in acidic solution,
direct reduction of CO_2 to methanol occurs, but that
formaldehyde is apparently bypassed as an intermediate
(6b). In basic solution there is no firm evidence as to
whether dissolved CO_2 or carbonate ion is reduced to
formate, or that formate is reduced to formaldehyde, even
at potentials more cathodic than -2.1V. Formaldehyde,
however, is reduced at moderate potentials to methanol,
although it has been reported that complications arise
from base promoted formation of polyoxymethylene glycols.
One conclusion to be drawn from Scheme 2 is the implied
kinetic difficulty of reducing carbonate or bicarbonate,
making acid solution, where CO_2 is the dominant form, the
preferred medium for CO_2 reduction.

Formation of oxalate in aqueous solution is a different matter. One brief report of CO_2 reduction in unbuffered water, where competitive water reduction increased the solution pH, gave evidence for oxalate production (6c). By inference, dimerization of CO_2 to oxalate is favored under conditions of high pH where the coupled H^+/e^- steps leading to formate, formic acid and methanol are relatively slow.

Recent work on non-traditional solid electrode surfaces, such as Ru coated carbon, (6q,r,u) or molybdenum, (6p) has been successful for reducing CO_2 to either methane or methanol. For example, with Ru coated carbon the reduction to methane depicted in Eq. 1 can be achieved with 24% Faradaic efficiency in acidic aqueous solution at ca. -0.38V versus NHE.

$$CO_2 + 8H^+ + \ ^-e^- \longrightarrow CH_4 + 2H_2O \qquad (1)$$

Likewise, on a molybedenum electrode methanol has been produced in >50% Faradaic efficiency at potentials between -0.7 and -0.8V vs. SCE. A single, but tantalizing report of both methane and ethylene production at copper electrodes has been claimed to occur between ca. -1.35V versus NHE (6t). The mechanisms by which these reductions occur, although of extreme importance in this field, are unknown.

In nonaqueous solvents, such as DMF or acetonitrile, electroreduction of CO_2 is followed by the three principal reduction pathways as depicted in Scheme 3. By using ultrafast sweep rate cyclic voltammetry Lamy, Nadjo and Saveant have determined the standard potential for the CO_2/CO_2^- couple to be -2.21 ± 0.015 V versus SCE in DMF with 0.1 M TEAP as supporting electrolyte (6g). The subsequent reactivity and distribution between formate, CO(g), or oxalate as final products is dependent upon factors such as the CO_2 concentration, the electrode material, the type of electrolyte, or the presence of adventitious acid (6,d,v).

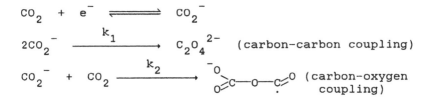

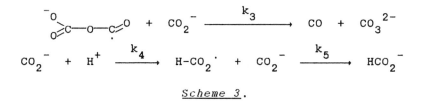

$$Scheme\ 3.$$

Amatore and Saveant have provided estimates of the rates for the various pathways shown in the scheme (**6d**). The fastest process is the carbon-carbon coupling process to yield oxalate, e.g., $10^7 \, M^{-1}sec^{-1}$, whereas both neutralization of CO_2^- by adventitious water and O,C coupling occur at ca. $10^3 \, M^{-1}sec^{-1}$. Of note is that the source of the second electron in these latter cases is predominantly from CO_2^- rather than diffusion to the electrode surface.

Our goals for the development of CO_2 electrocatalysts are twofold; the design of systems which operate at high turnover numbers near the thermodynamic potential for couples like CO_2/HCO_2H or $CO_2/H_2C_2O_2$, and, to understand mechanistic pathways well enough to design and synthesize new electrocatalysts that possess a high degree of product specificity, i.e., to be able to control the course of CO_2 reduction toward the formation of products such as formate, methanol or even methane rather than carbon monoxide.

PHOTOELECTROCHEMICAL REDUCTION OF CARBON DIOXIDE USING SEMICONDUCTOR ELECTRODES. Several different strategies for carbon dioxide reduction on semiconductors electrodes have been used to produce CO, formic acid, or even methanol (**14**). These include:

    1.) Use of p-type semiconductors under band gap irradiation to directly reduce CO_2 and its intermediates.

    2.) Use of p-type semiconductors to photoreduce known CO_2 reduction catalysts.

Direct photoassisted reduction of CO_2 to formic acid occurs with Zn doped-p-type GaP as a photocathode in aqueous phosphate buffer (pH 6.8) using 365nm light at a cell bias potential of -1.0V (SCE). It is significant that smaller amounts of both formaldehyde and <u>methanol</u> were also observed under these conditions (**14f**).

Chemically catalyzed reduction of CO_2 using soluble redox catalysts as electron acceptors from a p-type Si photocathode has been reported by Bradley and co-workers. For example, by using $[(Me_6[14]aneN_4)Ni^{II}]^{2+}$ (vide infra) in $CH_3CN/[n-Bu_4N][ClO_4]$ solution, CO_2 could be reduced to CO with 752nm light at an applied potential of -1.0V (SCE) (14a). A related, but more interesting approach is to chemically modify a semiconductor electrode with the catalytic species of interest. This approach has been taken by Cabrera and Abruna using p-WSe$_2$ modified with poly-$[Re(vbpy)(CO)_3Cl]$ to produce CO at -0.65V (SCE) under irradiation with a He/Ne laser (vide infra) (14k). Undoubtedly the approach embodied by this work will be fertile ground for photoelectrochemical cell research in the near future.

REACTIVITY OF TRANSITION METAL COMPLEXES TOWARD CO$_2$. The existence of a well-developed chemistry of transition metal-CO_2 complexes would aid in the design of electrocatalysts, but unfortunately this chemistry remains obscure. At least four areas of significance can be identified where more information would enhance our ability to define possible reduction pathways:
　　　1.) The study of CO_2 bonding modes and their characterisitic reactivity properties, especially as a function of the central metal and coordination number.
　　　2.) Interconversion of side-bound and carbon-bound CO_2.
　　　3.) Electrophilic and nucleophilic attack at the coordinated CO_2 ligand.
　　　4.) The redox properties of coordinated CO_2.
　　　Despite the absence of systematic information in these areas, there have been synthetic, structural, and reactivity studies which provide a useful background to the CO_2 reduction problem, as has been described in recent reviews (2).

Bonding Modes and Their Reactivities. Many metal complexes in low (e.g., Ni(0)) or intermediate (e.g. Ru(II)) oxidation states react with CO_2 in solution, or in the solid state, although the isolation and complete characterization of the resulting metal-CO_2 complexes is difficult. From the results of x-ray crystallography, seven different structural types have been identified:

1.)   Tetrahedral Ni(0); $(Cy_3P)_2Ni(\eta^2-CO_2)$, Cy is cyclohexyl (*7r,s*).

2.)   Octahedral Mo(0); $(PMe_3)_3(CNR)Mo(\eta^2-CO_2)_2$, where R is Me, i-Pr or t-Bu (*7n,q*).

3.) $Nb(\eta^5-C_5H_4Me)_2(CH_2SiMe_3)(\eta^2-CO_2)$ (*7p*).

4.)   Formal Mo(II); $[(\eta^5-C_5H_5)_2Mo(\eta^2-CO_2)]_n$ (*7v*).

5.)   Octahedral Rh(III); $diars)_2(Cl)Rh(\eta^1-CO_2)$, where diars is 1,2-(bis-dimethylasino)ethane (*71*).

6.) $[Co(Salen)(\mu-CO_2)(K)(THF)]_n$ (*7t*).

7.) $[((CO)_4Re)_2(\mu-CO_2)_2(Re(CO)_5)_2$ (*7b*).

The first four cases involve side-bound coordination which is reminiscent of metal-alkene complexs as shown in structure I, while cases 4-6 involve at least some carbon-metal interaction.

(I)

The carbon-bound complexes show more diverse structural behavior than the side-bound species, for example, the Rh complex is strictly monohapto with respect to the metal as shown in II, while the Co complex has both oxygen atoms "supported" by coordination to K^+.

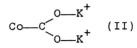

(II)

(where the K^+ ions are chemically inequivalent)

The Re cluster discovered by Beck and coworkers (*7b*) is similar in that Re^I is coordinated to both C and O in the manner depicted in III.

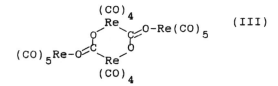

(III)

Although no systematic studies of the reactivity of complexed CO_2 have appeared, the dominant mode of reaction seems to be oxygen transfer to suitably acidic acceptors such as water. Thus, the attempted preparation of low-valent CO_2 complexes often leads to the isolation of metal-bicarbonato or metal-carbonyl species in a higher oxidation state, presumably due to reactions like those shown in Eqs. 2a and 2b. The released carbonate ion is a strong nucleophile in non-aqueous media and may attack metal ions in solution to form carbonate or bicarbonate complexes.

$$M(\eta-CO_2) + H_2O \longrightarrow [M(CO)]^{2+} + 2OH^- \qquad (2a)$$

$$2OH^- + 2CO_2 \longrightarrow 2CO_3^{2-} \qquad (2b)$$

An intramolecular example of CO_2 acting as a Lewis acid, oxygen acceptor, has been demonstrated by the x-ray crystal structure of the reaction product between $IrCl(C_8H_{14})(PMe_3)_3$ and CO_2 in benzene (71) The complex, $IrCl(C_2O_4)(PMe_3)_3$, contains the metallocycle shown in IV, which is an isomer of oxalate, and can be viewed as arising from an intramolecular acid-base interaction between CO_3^{2-} and CO.

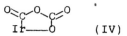 (IV)

Intermolecular examples are known in metal carbonyl-anion chemistry where reductive oxygen transfer occurs, apparently between free and complexed CO_2. An example is shown in Eq. 3, although the intermediate CO_2 complex is apparently too reactive to appear as an observable intermediate (8a).

$$Li_2[W(CO)_5] + 2CO_2 \longrightarrow W(CO)_6 + Li_2CO_3 \qquad (3)$$

Other oxygen acceptors have been identified, most notably PR_3, which results in the corresponding phosphine oxide and CO; a reaction type that could be exploited in future electrocatalytic cycles (8d).

Of greater potential interest is the reaction of coordinated CO_2 with with reactants that attack at carbon. Although there are few examples, the acrylic acid formation that results via ethylene-CO_2 coupling shown in Eq. 4 is noteworthy (7n).

trans-Mo$(\eta^2$-$C_2H_4)_2$(PMe$_3)_4$ + CO_2 ⟶

$$(4)$$

$$1/2[Mo(H_2C=CHCO_2H)(\eta^2-C_2H_4)(PMe_3)_2]_2$$

Whether the reaction proceeds in an inter- or intramolecular fashion is not known.

Protonation (or electrophilic attack) at carbon to give formate or formate precursors is another possible reaction if η^2-CO_2 exhibits reactivity modes similar to dihapto alkenes. Later we will discuss recent kinetic evidence of this type of reactivity in the electrocatalytic reduction of CO_2 by [Os(bpy)$_2$(CO)H]$^+$.

Another important reaction is the coupling of two carbon dioxide molecules at the carbon atoms to give oxalate. This process appears to be a side reaction in some electrocatalytic reductions (3j,p,1,w) yet it has never been observed as a reaction pathway from CO_2 complexes. However, a model of the reaction exists in the example of the carbon-carbon coupling of dimethyl malonate by $(\eta^5$-$C_5H_5)_2$Ti(CO)$_2$ (9).

<u>Insertion of CO_2 into Metal-Ligand Bonds</u>. Examples of synthetic reactions where CO_2 inserts into metal hydride, alkyl, aryl, alkoxide, hydroxide, and amide bonds are well known (10). Only recently have kinetic and mechanism studies been conducted which reveal the details of the insertion process on the molecular level. Notable is the work of Darensbourg and coworkers (10q,1,o) on the W-alkyl insertion shown in Eq. 5:

$$[(PR_3)(CO)_4WMe]^- + CO_2 \xrightarrow{\text{THF}} [(PR_3)(CO)_4W-O\overset{O}{\overset{\|}{C}}-Me]^- \quad (5)$$

Recently we have reported (10q,r) the first detailed kinetic studies of CO_2 insertion into both a metal hydride bond and a related metal alkoxide bond, i.e., that of <u>fac</u>- Re(bpy)(CO)$_3$H (Eq. 6), and

<u>fac</u>-Re(bpy)(CO)$_3$O$\overset{Ph}{\underset{Me}{C}}$-H (Eq. 7).

$$\underline{fac}\text{-Re(bpy)(CO)}_3\text{H} + \text{CO}_2 \xrightarrow{\text{THF}} \underline{fac}\text{-Re(bpy)(CO)}_3\text{O}_2\text{CH} \qquad (6)$$

$$\underline{fac}\text{-Re(bpy)(CO)}_3\text{O}\overset{\text{Ph}}{\underset{\text{Me}}{\text{C}}}\text{-H} + \text{CO}_2 \xrightarrow{\text{CH}_3\text{CN}}$$

$$(7)$$

$$\underline{fac}\text{-Re(bpy)(CO)}_3\text{O}-\overset{\text{O}}{\text{C}}-\text{O}\overset{\text{Ph}}{\underset{\text{Me}}{\text{C}}}\text{-H}$$

      Both the hydride and alkoxide exhibit psuedo-first order kinetics in the presence of ca. 10^{-3}M to 0.33M CO_2 in solvents such as THF, acetone, or CH_3CN. Figure 1 shows typical decay kinetics for $\underline{fac}$-Re(bpy)(CO)$_3$O$\overset{\text{Ph}}{\underset{\text{Me}}{\text{C}}}$-H in THF solution; the spectral changes exhibit isobestic behavior at 346 nm and 411 nm that corresponds to the decay and appearance, respectively, of the metal-to-ligand charge transfer absorption bands of the reactant alkoxide and the product metallocarboxyester complexes. Concentration studies establish a first order dependence for CO_2, and the simple rate law, $-d[\text{Re}]/dt=k_i[\text{Re}][\text{CO}_2]$, holds for both insertions.

      The temperature dependencies support an associative process for both insertions (for Re-H; $\Delta H^\dagger = 12.8$ kcal/mole, $\Delta S^\dagger = -33.0$ eu; for Re-OCH(Me)Ph; $\Delta H^\dagger = 10.9$ kcal/mole, $\Delta S^\dagger = -27.6$ eu) and the activation parameters compare favorably with the case of $[(\text{MeO})_3\text{P(CO)}_4\text{WMe})]^-$ ($\Delta H^\dagger = 10.2$ kcal/mole, $\Delta S^\dagger = -43.3$ eu).

      That significant charge transfer character is involved in both mechanisms is indicated by the large, general dielectric effect on the insertion rate, a phenomenon that is larger for the hydride reaction than the alkoxide. For example, the second order rate constant, k_i, for $\underline{fac}$-Re(bpy)(CO)$_3$H in THF is 1.97×10^{-4} $\text{M}^{-1}\text{sec}^{-1}$ but in CH_3CN is 5.44×10^{-2} $\text{M}^{-1}\text{sec}^{-1}$, while for $\underline{fac}$-Re(bpy)(CO)$_3$O-$\overset{\text{Ph}}{\underset{\text{Me}}{\text{C}}}$-H in THF the rate is 2.2×10^{-2} $\text{M}^{-1}\text{sec}^{-1}$ and in CH_3CN it is 1.62×10^{-1} $\text{M}^{-1}\text{sec}^{-1}$.

      One possible mechanism for the metal-alkoxide insertion reaction is a water or alcohol catalyzed chain

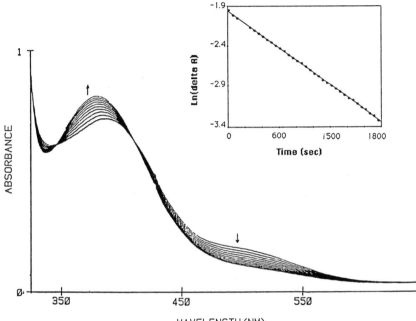

FIGURE 1. Spectral changes which occur in the visible region of the spectrum upon insertion of CO_2 into the metal-alkoxide bond of <u>fac</u>-Re(bpy)(CO)$_3$OC(H)(Me)Ph. Inset is typical pseudo-first order decay kinetics monitored at 450nm in THF solution.

like that proposed by Chisolm and coworkers for amide and alkoxide complexes of the mid- and early transition elements (Eqs. 8-9) (10s,t).

$$R-OR \; + \; CO_2 \; \rightleftharpoons \; ROCO_2H \qquad (8)$$
$$[R \text{ is aryl or alkyl}]$$

$$M-OR' \; + \; ROCO_2H \; \longrightarrow \; M-O-\overset{\displaystyle O}{\overset{\|}{C}}-OR \; + \; R'OH \qquad (9)$$

To rule out this possibility we have followed the course of the insertion reaction in the presence of D_2O using ^1H Nmr spectroscopy. Figure 2a shows the proton spectrum of fac-Re(bpy)(CO)$_3$O$\overset{\displaystyle Ph}{\underset{\displaystyle Me}{C}}$-H in CD_3CN emphasizing the assignments for the bpy and phenethylalkoxide groups. The spectrum shows that the lack of symmetry at the alkoxide carbon is dramatically felt at the bpy. Addition of 1.36M D_2O to the solution causes only slight shifts, presumably due to specific solvation effects like hydrogen bonding (Fig. 2b). From this result we can be sure that hydrolysis (Eq. 10) does not occur on the timescale required to obtain the spectrum, i.e., ca. 30 min, since any substitution process that produces the thermodynamically favored facial isomer would result in formation of a Re complex with a symmetry plane.

$$\text{fac-Re(bpy)(CO)}_3\text{O}\overset{\displaystyle Ph}{\underset{\displaystyle Me}{C}}\text{-H} \; + \; D_2O \; \longrightarrow \; \text{fac-Re(bpy)(CO)}_3\text{OD} \; +$$

$$(10)$$

$$\text{DO}\overset{\displaystyle Ph}{\underset{\displaystyle Me}{C}}\text{-H}$$

Figure 2c shows the same experiment in CO_2 saturated solution where it is clear that fac-Re(bpy)(CO)$_3$O-$\overset{\displaystyle O}{\overset{\|}{C}}$-O$\overset{\displaystyle Ph}{\underset{\displaystyle Me}{C}}$-H is cleanly formed. Note that the spectrum of the bpy region still shows the effect of the chiral carbon of the metallocarboxyester, although it is attenuated by the greater distance between the bpy ligand and the chiral center.

From the data at hand it is possible to propose a common type of charge-separated state for both hydride and alkoxide insertions; these are depicted in structures V and VI.

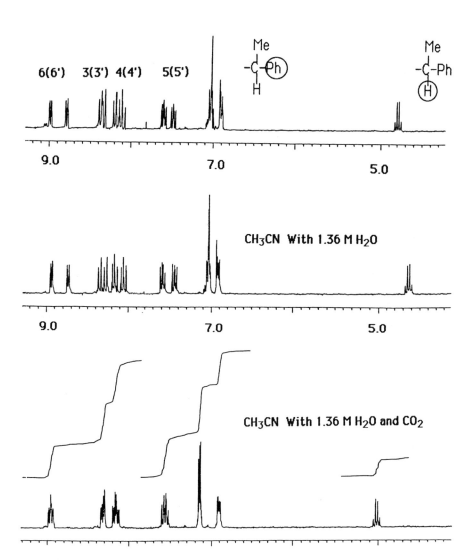

FIGURE 2. Proton NMR spectral experiments
demonstrating the hydrolytic stability of
fac-Re(bpy)(CO)$_3$OC(H)(Me)Ph in CD$_3$CN solution. Shifts
are realtive to TMS as an external standard; see text
for explanation of the spectral changes.

(V)   (VI)

        Involvement of V as the transition state of the
hydride reaction is further supported by the appearance
of an inverse isotope effect of ca. 1/2 (e.g., for CH_3CN,

k_H/k_D = 0.53), and by the correlation of σ_p^+ and k_i for a
series of 4, 4'-disubstituted bpy complexes.
        The associative, alkoxide attack on the CO_2 carbon
depicted in VI bears formal resemblance to that of V,
however, an alternative mechanism is a carbonium ion
migration from the alkoxide oxygen to the CO_2 carbon.
This interesting alternative is the concern of some of
our current studies.
        Photochemical Reduction of CO_2. The reaction of CO_2
with fac-Re(bpy)(CO)$_3$H, as shown in Eq. 6, is also
photocatalyzed in solvents like THF or benzene where the
thermal insertion is inherently slow (12b). Another
example of a photochemically driven insertion is that of
Re(diphos)$_2H_3$ (diphos is $Ph_2PCH_2CH_2PPh_2$) which eliminates
H_2 on UV irradiation to generate the coordinately
unsaturated complex Re(diphos)$_2$H (or a solvated form)
that reacts with CO_2 to form Re(diphos)$_2(\eta^2-O_2CH)$ (12c).
        Another approach has been taken by Ziessel, Lehn and
coworkwers, who in a series papers, both interesting and
rich, use the reductive quenching of Ru(bpy)$_3^{2+}$ and
Re(bpy)(CO)$_3$Cl to produce formate and CO, respectively
(12a,d,e).

ELECTROCATALYTIC REDUCTION OF CO_2 IN HOMOGENEOUS
SOLUTION.

There is very little mechanistic information on the
electrocatalytic reduction of CO_2. In many cases either
the the electrochemical events or the products of the
reduction are not known. Briefly, the few
electrocatalysts can be grouped into the following
structural categories:
        1.) Complexes of porphyrins, phthalocyanines, and
related macrocylic ligands.

2.)   Other square planar complexes.
3.)   Metal clusters.
4.)   Metal-polypyridine complexes.

In the following discussion the mechanistic findings and the implications of the little available work is presented, and then it is contrasted with our own recent studies.

METAL PORPHYRINS, PHTHALOCYANINES, AND RELATED MACROCYCLES. Other than the metal-polypyridine complexes, macrocyclic metal complexes have provided the largest number of catalysts. The phenomenological aspects of this chemistry have been investigated by Fisher and Eisenberg (3h), Lieber and Lewis (3n), Sauvage and coworkers (3e), and recently by Becker and coworkers (3l). The only study which has attempted to address mechanistic questions in detail is that of Pearce and Pletcher (3u), although both the work of Sauvage et. al. and Fisher and Eisenberg nicely demonstrate that mechanistic selectivity for CO_2 reduction versus H_2O reduction exists.

In their study, Pearce and Pletcher examined some of the original electrocatalysts discovered by Fisher and Eisenberg such as Co^{II}(Salen) and $Ni(teta)^{2+}$ in a mixed CH_3CN/H_2O solvent. They identified bicarbonate as the oxygen containing by-product of the reaction, and postulated that H_2O is the oxygen acceptor. From cyclic voltammetry and bulk electrolysis studies the following mechanistic sequence was then proposed:

$$[M^{II}L]^{n+} + e^- \rightleftharpoons [M^I L]^{(n-1)+} \tag{11}$$

$$[M^I L]^{(n-1)+} + CO_2 + Na^+ \rightleftharpoons [M^{III}LCO_2^- Na^+]^{n+} \tag{12}$$

$$[M^{III}LCO_2^- Na^+]^{n+} + H_2O \xrightarrow{fast} [M^{III}LCOOH]^{n+} + NaOH \tag{13}$$

$$[M^{III}LCOOH]^{n+} + e^- \longrightarrow [M^{II}LCOOH]^{(n-1)+} \tag{14}$$

$$[M^{II}LCOOH]^{(n-1)+} \xrightarrow{slow} [M^{II}L]^{n+} + CO + {}^-OH \tag{15}$$

$$2NaOH + 2CO_2 \longrightarrow 2NaHCO_3 \tag{16}$$

A direct model of the CO_2 complex in Eq. 12 is offered by the x-ray structure of $[Co(pr-salen)K(CO_2)THF]_n$ which shows a C-bound CO_2 with the oxygen coordinated to K^+ (7t).

Employing a very different strategy, Kapusta and Hackerman (3k) have examined the electrocatalytic behavior of Co-phthalocyanine films deposited on carbon electrodes and find in aqueous solution, over a wide pH range, good current yields of formate. At low pH values, however, up to 5% methanol is produced. If confirmed these results imply that the multielectron, stepwise reduction of CO_2 to methanol via metal electrocatalysts ia a viable strategy.

OTHER SQUARE-PLANAR COMPLEXES. The majority of the macrocylic systems, like those discussed above, are square planar and contain neutral or negatively charged nitrogen or mixed nitrogen and oxygen donor atoms. An electronically and sterically distinct class of complex is represented by $Rh(diphos)_2^+$, which has recently been shown to reduce CO_2 to formate. Under electrocatalytic conditions, Slater and Wagenknecht (3g) have suggested the intermediate formation of $Rh(diphos)_2H$ by radical abstraction from the CH_3CN solvent is the first step in the reduction, i.e., Eqs. 17 and 18.

$$[Rh^I(diphos)_2]^+ + e^- \rightleftharpoons [Rh(diphos)_2]^0 \qquad (17)$$

$$[Rh(diphos)_2]^0 + CH_3CN \longrightarrow Rh(diphos)_2H + {}^{\cdot}CH_2CN \quad (18)$$

In their proposed mechanism, subsequent insertion of CO_2 into the Rh-H bond and dissociation of product formate completes the catalytic cycle.

METAL CLUSTERS. Although as a class clusters are promising as electrocatlysts only two relevant studies have appeared, one concerning the electroreduction of Fe-S clusters which reduce CO_2 to a variety of products including CO and formate (3j), and the other involving the catalytic chemical reduction of CO_2 to CO by Ru carbonyl clusters (8e) such as $Ru_4(CO)_{12}^{4-}$. No mechanisitic information is available for either of these systems.

METAL-POLYPYRIDINE COMPLEXES. The majority of the mechanistic data has come from studies on complexes containing polypyridyl ligands. Among the appealing properties of ligands like 2, 2'-bipyridine (bpy) and 1, 10-phenanthroline (phen) is that they stabilize metals in a large number of oxidation states while at the same time they are "electron reservoirs" capable of storing electrons at potentials between ca. -0.7 and -1.7V by utilizing vacant π^* orbitals.

Cobalt Bipyridine Complexes. The work of Keene, Cruetz, and Sutin (8f) on the stoichiometric reduction of CO_2 by $[Co^I(bpy)_3]^+$ in buffered aqueous solutions demonstrates that two electrons, both of which are metal-based, can be used to reduce CO_2 to CO and in this sense is similar to CO_2 reduction with Co-macrocycles. Ambiguities remain concerning the mechanism, especially whether CO_2 or HCO_3^- is the substrate. The end result of their kinetic studies is consistent with either $Co(bpy)_2(H_2O)H^{2+}$ reacting with HCO_3^-, or $Co(bpy)_2(H_2O)_2^+$ with CO_2, although they favor the former interpretation. Of particular interest is the deactivation pathway shown in Eq. 19 where the CO product intercepts the Co-bpy reagent precipitating a Co dimer.

$$Co(bpy)_3^+ + 2CO \longrightarrow 1/2[Co(bpy)(CO)_2]_2 + 2bpy \qquad (19)$$

Ru(trpy)(dppene)L^{n+} Complexes (trpy is 2,2,'2"-terpyridine; dppene is 1,2-bis(diphenylphosphino)ethylene; L is Cl−,n=1, CH$_3$CN and CO,n=2). Based on the results of cyclic voltammetry and bulk electrolysis studies we have found that the above complexes complexes undergo a two electron reduction process at potentials between −1.06 and −1.30V (versus SCE) to generate a highly reactive reduced intermediate, Ru(trpy)(dppene), which while not isolable, is reactive toward CO_2 to give CO (3c). From the results of chemical reactivity and cyclic voltammetry (see Fig. 3) the mechanism shown in Eqs. 20-25 can be proposed to account for the formation of CO in this system. Although the various steps in the mechanism are reasonable based on the proposed chemistry and the obseved products, direct evidence is available for only the steps shown in Eqs. 20 and 25. The equilibrium shown in Eq. 24, however, has been inferred from chemical studies.

$$[Ru(trpy)(dppene)L]^{n+} + 2e^- \xrightarrow{\quad CH_3CN \quad} Ru(trpy)(dppene) + 2[L]^{n+} \qquad (20)$$

$$Ru(trpy)(dppene) + CO_2 \longrightarrow Ru(trpy)(dppene)(CO_2) \qquad (21)$$

Ru(trpy)(dppene)(CO$_2$) + [N(n-Bu)$_4$]$^+$ $\longrightarrow$

[(typy)(dppene)Ru-$\overset{\overset{\text{O}}{\|}}{\text{C}}$-OH]$^+$ + N(n-Bu)$_3$ + CH$_3$CH$_2$CH=CH$_2$ (22)

[(trpy)(dppene)Ru-$\overset{\overset{\text{O}}{\|}}{\text{C}}$-OH] $\rightleftharpoons$

[(trpy)(dppene)Ru(CO)]$^{2+}$ + $^-$OH   (23)

[(trpy)(dppene)Ru(CO)]$^{2+}$ + 2e$^-$ $\longrightarrow$
Ru(trpy)(dppene) + CO   (24)

$^-$OH + CO$_2$ $\longrightarrow$ HO$\overset{\overset{\text{O}}{\|}}{\text{C}}$-O$^-$ (25)

The Ru-trpy system can only achieve rates of 0.15 turnovers/min with 70-90% Faradaic efficiency at a potential of between -1.3 and -1.4V using a Pt gauze electrode. Catalyst deactivation occurs slowly in the presence of CO$_2$, but experiments in the absence of CO$_2$ show the rapid decomposition of Ru(trpy)(dppene). Careful inert atmosphere experiments show that this putative intermediate is not isolable using our present techniques.

Several important mechanistic points have emerged from the above studies. First, complexes exhibiting

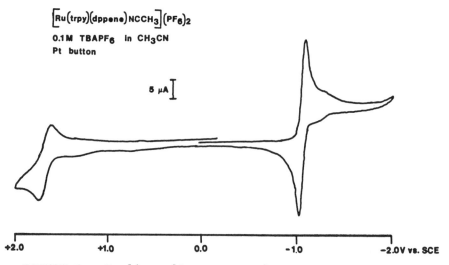

[Ru(trpy)(dppene)NCCH$_3$](PF$_6$)$_2$
0.1M TBAPF$_6$ in CH$_3$CN
Pt button

5 μA

+2.0   +1.0   0.0   -1.0   -2.0V vs. SCE

FIGURE 3. Cyclic voltammogram of
Ru(trpy)(dppene)(CH$_3$CN)$^+$ showing the near simultaneous
two electron reduction process characteristic of this type of complex.

closely spaced metal-based (i.e. Ru(II)/Ru(I)) and
ligand-based (trpy/trpy(-)) couples can be prepared and
their reducing equivalents transferred in virtually a two
electron fashion. And second, the proposed CO_2 complex
shown in Eq. 22 must be extremely basic since it can
effect the Hofmann degradation of $N(n-bu)_4^+$ under mild
conditions.

<u>Rh(bpy)$_2$X$_2^+$ Complexes (bpy is 2,2'-bipyridine and X</u>
<u>is Cl$^-$ or $^-$O-$\overset{O}{\underset{}{S}}$-CF$_3$)</u>. Our work on 2, 2'- bpy complexes of
Rh has demonstrated methods for the catalytic reduction
of CO_2 to formate (<u>3c</u>). Figure 4 and Eqs. 26-31 portray
a series of molecular steps that are consistent with our
chemical and electrochemical results. The steps in Eqs.
26-28 are based on electrochemical results of Hanck and

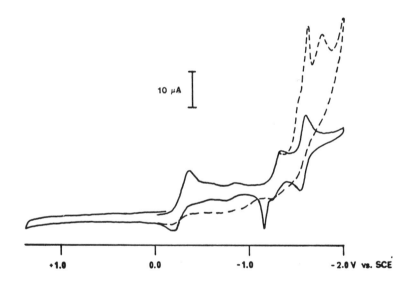

FIGURE 4. Cyclic voltammogram of
[<u>cis</u>-Rh(bpy)$_2$(O$_3$SCF$_3$)$_2$]$^+$ taken in CH$_3$CN/0.1M TBAH with
a glassy carbon button working electrode. The most
positive reduction wave corresponds to a two electron
process coupled to loss of Cl$^-$ and formation of
[Rh^I(bpy)$_2$]$^+$. The two sequential one elelectron
reductions are bpy-based processes involving
[Rh^I(bpy)$_2$]$^+$.

DeArmond and their coworkers (11). Furthermore, the
existence of an intermediate CO_2 complex in Eq. 29 is

$$[Rh(bpy)_2X_2]^+ + 2e^- \xrightarrow{\ CH_3CN\ } [Rh(bpy)_2]^+ + 2X^- \qquad (26)$$

$$[Rh(bpy)_2]^+ + e^- \rightleftharpoons [Rh^I(bpy)(b\bar{p}y)] \qquad (27)$$

$$[Rh^I(bpy)(b\bar{p}y)] \rightleftharpoons [Rh(b\bar{p}y)_2]^- \qquad (28)$$

$$[Rh(b\bar{p}y)_2]^- + CO_2 \longrightarrow [Rh(bpy)_2(CO_2)]^- \qquad (29)$$

$$[Rh(bpy)_2(CO_2)]^- + [N(n\text{-bu})_4]^+ \longrightarrow Rh(bpy)_2(CO_2H) +$$
$$N(n\text{-bu})_3 + H_2C=CHCH_2CH_3 \qquad (30)$$

$$[Rh(bpy)_2(CO_2H)] + 2e^- \longrightarrow [Rh(bpy)_2]^+ + H\overset{O}{\underset{}{C}}\text{-}O^- \qquad (31)$$

strongly implicated by the fact that <u>no</u> reaction occurs
between the multiple-reduced species $[Rh(b\bar{p}y)_2]^-$ and

$N(n\text{-bu})_4^+$.

For these catalysts, formate production is found to
occur at a rate of ca. 0.2 turnovers/min at -1.55V using
a carbon cloth electrode with an initial current
efficiency of >80%. The system slowly degrades by two
routes, one which leads to the production of an
intermediate that is catalytic toward H_2 evolution from
the medium, and the other which results in the deposition
of an insoluble complex. Neither of the decomposition
processes have been studied although the H_2 producing
reaction is intriguing since it apparently involves a
Hofmann degradation pathway that gives H_2 at the expense
of quaternary ammonium salts.

The studies of the Rh catalysts reveal that bpy/bpy$^-$
redox couples can act as internal electron transfer sites
for the ultimate delivery of two electrons to a
coordinated CO_2 molecule. In addition, the results show
that a feeble acid like $N(n\text{-bu})_4^+$ can act as the oxygen
sink for CO formation (like the Ru-trpy systems) and, in
addition, can effectively act as a proton source for
formate production.

Rhenium Polypyridine Complexes. The early studies
of Lehn and coworkers (3m) described the formation of CO
and trace formate from the bulk electrolysis of
<u>fac</u>-Re(bpy)(CO)$_3$X (X is Br or Cl) in CO_2 saturated DMF.

These observations have been confirmed by us (**3b**), and by Breikss and Abruna (**3t**). For fac-Re(bpy)(CO)$_3$Cl, Ziessel (**1g**) has suggested a mechanism for CO production that appears to invoke a contra-thermodynamic step, i.e., the production of the strong oxidant Re^{II}(bpy)(CO)$_3$Cl$^+$ (E$_{1/2}$ ca. +1.3V) as shown below in Eqs. 31-33.

$$Re(bpy)(CO)_3Cl + e^- \xrightleftharpoons{\quad E_{1/2} = -1.35V \quad} [Re(b\bar{p}y)(CO)_3Cl]^- \quad (31)$$

$$[Re(bpy)(CO)_3Cl]^- + CO_2 + 2H^+ \longrightarrow [Re^{II}(bpy)(CO)_3Cl]^+ + CO + H_2O \quad (32)$$

$$[Re^{II}(bpy)(CO)_3Cl]^+ + 2e^- \longrightarrow Re(bpy)(CO)_3Cl \quad (33)$$

In this sequence an intermediate that involves a sesqui-bpy ligand, that is, a bipyridine with one arm not coordinated to the Re, has been suggested. Such intermediates have been postulated in substitution reactions of metal-bpy complexes, but would be expected to lead to rapid bpy loss and subsequent, rapid, loss of catalytic activity for the system. Future experiments should be designed to illuminate this somewhat counterintuitive suggestion.

From the results of our work with the Re system CO or formate is formed from four competitive pathways. Also, another pathway must exist which is responsible for the production of small amounts of oxalate. Eqs. 34-39 represent the series of steps which we believe are responsible for the production of CO from highly reducing, electrogenerated Re(bpy)(CO)$_3$ radicals (or the solvated form Re(bpy)(CO)$_3$(CH$_3$CN)).

$$fac\text{-}Re(bpy)(CO)_3X + e^- \xrightleftharpoons{\quad CH_3CN \quad} [Re(bpy)(CO)_3X]^- \quad (34)$$

$$[Re(bpy)(CO)_3X]^- \xrightarrow{\quad slow \quad} Re(bpy)(CO)_3 + X^- \quad (35)$$

$$Re(bpy)(CO)_3 + CO_2 \rightleftharpoons Re(bpy)(CO)_3(CO_2) \quad (36)$$

$$2Re(bpy)(CO)_3(CO_2) \longrightarrow$$
$$(CO)_3(bpy)Re\text{-}O\text{-}\overset{\overset{\displaystyle O}{\|}}{C}\text{-}O\text{-}Re(bpy)(CO)_3 + CO \quad (37)$$

$$(CO)_3(bpy)Re-O-\overset{\overset{O}{\|}}{C}-O-Re(bpy)(CO)_3 \ + \ e^- \longrightarrow$$

$$Re(bpy)(CO)_3 \ + \ ^-O\overset{\overset{O}{\|}}{C}O-Re(bpy)(CO)_3 \quad (38)$$

$$Re(bpy)(CO)_3CO_3^- \ + \ e^- \longrightarrow Re(bpy)(CO)_3 \ + \ CO_3^{2-} \quad (39)$$

In the above scheme, the existence of the intermediate
CO_2 complex can be inferred from competition studies, and
from the existence of a carbonato bridged dimer <u>analog</u>,
$(CO)_3(bpy)Re-O-\overset{\overset{O}{\|}}{C}-O-Re(bpy)(CO)_3$, which has recently been
detected by *in situ* FT-IR methods following the
photolysis of the substituted Re dimer shown in Eq. 40.

$$[Re(bpy')(CO)_3]_2 \ + \ 2CO_2 \ \xrightarrow[\lambda>450nm]{THF, \ 25 \ C}$$

$$(bpy')(CO)_3Re O\overset{\overset{O}{\|}}{C}ORe(CO)_3(bpy') \ + \ CO \quad (40)$$

(where bpy' is (4,4'-di-<u>tert</u>-butyl)-2,2'-bipyridine)

A non-destructive deactivation route for this
pathway involves the formation and precipitation of the
bicarbonate complex shown in Eq 41, where the source of
protons is either adventitious, or deliberately added
H_2O.

$$\underline{fac}-Re(bpy)(CO)_3CO_3^- \ + \ H_2O \longrightarrow$$

$$\underline{fac}-Re(bpy)(CO)_3CO_3H \ + \ OH^- \quad (41)$$

The novelty of the Re-bpy catalytic system is
further demonstrated by the involvement of a two-electron
pathway based on the red-purple anion $Re(bpy)(CO)_3^-$.
This exceedingly reactive species can be generated either
from a second reduction of fac-$Re(bpy)(CO)_3X$ complexes,
or from the metal-metal bonded dimeric species
$[fac$-$Re(bpy)(CO)_3]_2$ (Eqs. 42 and 43).

$$[Re(bpy)(CO)_3X]^- \ + \ e^- \ \xrightarrow[fast]{CH_3CN} \ [Re(bpy)(CO)_3]^- \ + \ X^- \quad (42)$$

$$[fac\text{-}Re(bpy)(CO)_3]_2 + 2e^- \xrightarrow{\text{DMF}} 2[Re(bpy)(CO)_3]^- \quad (43)$$

Like the Re radical, the anion drives the electrocatalytic reduction of CO$_2$ to CO in CH$_3$CN solution, but because of instability of the catalytic system at the high potentials necessary (-1.7 to -1.8V), and the rapid decomposition of the anion, only a tentative reduction mechanism can be proposed (See Eqs. 44-45).

$$Re(bpy)(CO)_3^- + CO_2 \xrightarrow{\text{CH}_3\text{CN}} [Re(bpy)(CO)_3(CO_2)]^- \quad (44)$$

$$Re(bpy)(CO)_3(CO_2)^- + CO_2 + 2e^- + [A] \longrightarrow$$
$$[Re(bpy)(CO)_3]^- + CO + [A\text{-}O]^- \quad (45)$$

The nature of the oxygen acceptor, A, is unknown, but in recent experiments with fac-[Re(bpy)(CO)$_3$(CH$_3$CN)]$^+$ we believe that the anion is acessible at lower potentials from other fac-[Re(bpy)(CO)$_3$L]$^{n+}$ derivatives and therefore intend to characterize this pathway more completely in later studies.

A third mechanistic path which leads to the production of formate appears to arise from the insertion of CO$_2$ into the metal-hydride bond of fac-Re(bpy)(CO)$_3$H (Eq. 46).

$$fac\text{-}Re(bpy)(CO)_3H + CO_2 \xrightarrow{\text{CH}_3\text{CN}} fac\text{-}Re(bpy)(CO)_3CO_2H \quad (46)$$

Under our conditions of bulk electrolysis the formate complex can be labilized, as in the chloro case, by either one or two electron reduction as shown in Eqs. 47-49, thus completing a catalytic cycle capable of producing free formate.

$$Re(bpy)(CO)_3CO_2H + e^- \longrightarrow [Re(bpy)(CO)_3O_2CH]^- \quad (47)$$

$$[Re(bpy)(CO)_3OC_2H]^- \xrightarrow{\text{slow}} Re(bpy)(CO)_3 + {}^-O_2CH \quad (48)$$

$$[Re(bpy)(CO)_3OC_2H]^- + e^- \xrightarrow{\text{fast}}$$
$$[Re(bpy)(CO)_3]^- + {}^-O_2CH \quad (49)$$

Our recent detailed kinetic study (vide supra) on the intimate mechanism of CO_2 insertion shown in Eq. 47 demonstates that it is an associative process exhibiting a high degree of bond breaking of the Re-H bond concurrent with bond forming between the hydride ligand and the carbon of the CO_2. The importance of this finding is that the rest of the Re coordination sphere is inert during the formal hydride transfer process. This observation carries with it the implication that such a mechanism could lead to electrocatalyst stability during the insertion act. A question which we are attempting to answer at the present is which of the available CO_2 reduction pathways is dominant at a given electrolysis potential. As discussed above, the possiblities include the radical "one-electron", the anion "two-electron" or, the hydride insertion pathways. There is the liklihood, however, of a _fourth_ pathway involving the direct interaction of CO_2 with $[Re(bpy)(CO)_3X]^-$. The only evidence of this has been provided by the flash photolysis studies of Kutal and coworkers (_12g_).

Currently, we are investigating the cyclic voltammetry of _fac_-Re(bpy)(CO)$_3$Br in an attempt to determine the relative contributions of these various pathways (_13_). Fig 5 shows a cyclic voltammogram of the complex in the presence and absence of CO_2. Perhaps the most startling finding is the direct evidence that $[fac-Re(bpy)(CO)_3Br]^-$ is only a catalyst precursor (Fig. 5b), and that the rate limiting step is the further reduction of an intermediate complex. We are currently probing the events that occur between the formation of the reduced complex and the formation of the intermediate by monitoring the disappearance of the intial complex by single-sweep techniques. Thus far, the following points have been established concerning the disappearance of $[Re(bpy)(CO)_3Br]^-$:

1.) In the absence of CO_2 (in TBAH/CH$_3$CN solution) both Re(bpy)(CO)$_3$(CH$_3$CN) and $[\underline{fac}-Re(bpy)(CO)_3]_2$ are formed and the addition of TBABr suppresses the formation of both.

2.) In the presence of CO_2 (0.14M in TBAH/CH$_3$CN solution) Re(bpy)(CO)$_3$(CH$_3$CN) and the expected radical coupling product $[fac-Re(bpy)(CO)_3]_2$ are not formed but the rate of disappearance of the $[Re(bpy)(CO)_3Br]^-$ is enhanced. Addition of Br^- however, decreases the rate of disappearance.

3.)   In TBAH/CH$_3$CN solution with 0.14M CO$_2$

increasing the concentration of anion [Re(bpy)(CO)$_3$Br]$^-$

increases its disappearance rate.

4.)   At the potentials necessary to produce

[Re(bpy)(CO)$_3$Br]$^-$, in the presence of excess Br$^-$, rapid

electron transfer-catalyzed substitution of Br$^-$ for
solvent occurs, most likely by the route shown in Eqs.
50-51:

$$Re(bpy)(CO)_3(CH_3CN) \; + \; Br^- \; \underset{\text{fast}}{\rightleftharpoons}$$
$$[Re(bpy)(CO)_3Br]^- \; + \; CH_3CN \quad (50)$$

$$[Re(bpy)(CO)_3Br]^- \; + \; [Re(bpy)(CO)_3(CH_3CN)]^+ \; \rightleftharpoons$$
$$\underline{fac}\text{-}Re(bpy)(CO)_3Br \; + \; Re(bpy)(CO)_3(CH_3CN) \quad (51)$$

Our data appears to show both a dissociative X$^-$ loss
pathway and an associative pathway where CO$_2$ reacts

directly with Re(bpy)(CO)$_3$X$^-$. From the rapid scan data
it appears that both routes result in the production of
the intermediate shown in Fig. 5b.

Bulk electrolysis of $\underline{fac}$-Re(bpy)(CO)$_3$Cl at -1.5V in

CH$_3$CN/TBAH medium using either a carbon cloth or Pt guaze

electrode gives CO in 92-99% current yield. The activity
at this potential can be up to several turnovers/min
although the long term stability has not been
tested. However, the system has been operated between
100-1000 turnovers without loss of catalytic activity.

Bulk electrolysis at -1.8V using the same conditions
as above gives CO in 80-90% current yield with rates in
the 1-10 turnover/min range, although in this case,
catalyst deactivation occurs rapidly, typically within
20-40 turnovers. In both cases the main deactivation
pathway appears to be the precipitation of

$\underline{fac}$-Re(bpy)(CO)$_3$O-$\overset{\text{O}}{\overset{\|}{\text{C}}}$-OH from the reaction mixture.

The Re system has led to several, possibly general
conclusions concerning the design of future catalytic
reactions:

1.) <u>One</u> electron reduction of a catalyst precursor
can lead to efficient net <u>two</u> electron reduction of CO$_2$.

2.) Electrochemical general of metal hydride
complexes which are capable of inserting CO$_2$ to give

formate via associative mechanisms could be viable, high
stability catalytic routes.

3.) A multiplicity of finely balanced pathways can
co-exist for a catalyst precursor some of which can
result in the same net chemistry, such as the reduction
of CO_2 to CO.

4.) A catalyst deactivation pathway can be
precipitation of highly polar, insoluble bicarbonate
complexes.

<u>$Os(bpy)_2(CO)H^+$</u>. Electrocatalytic reduction of CO_2
to give substantial yields of both CO and formate has
been achieved with this complex, and because of its
stability, mechanistic studies have provided the
unprecedented opportunity to explore what factors
determine formate formation at the expense of CO (<u>14w</u>).

In Fig. 6 is shown a series of cyclic voltammograms
which demonstrate that the catalytic properties of the
the complex are due to chemistry that originates from the
second bpy-based reduction wave. Using bulk electrolysis
and cyclic voltammetry techniques combined with digital
simulation methods, the following mechanism can be
proposed for electrocatalytic CO production in CH_3CN
solution using TBAH as supporting electrolyte at Pt or C
electrode surfaces:

$$[Os(bpy)_2(CO)H]^+ + e^- \underset{-1.35V}{\overset{CH_3CN}{\rightleftharpoons}} Os(bpy)(\bar{bpy})(CO)H \qquad (52)$$

$$Os(bpy)(\bar{bpy})(CO)H + e^- \underset{-1.55V}{\rightleftharpoons} [Os(\bar{bpy})_2(CO)H]^- \qquad (53)$$

$$[Os(\bar{bpy})_2(CO)H]^- + CO_2 \xrightarrow{k_1 = 40 \ M^{-1}s^{-1}} [I_{CO_2}]^- \qquad (54)$$

$$[I_{CO_2}]^- + CO_2 + e^- \xrightarrow{fast} [Os(bpy)(\bar{bpy})(CO)H]^+$$

$$CO + CO_3^{2-} \qquad (55)$$

$$2[I_{CO_2}]^- \xrightarrow{fast} 2Os(bpy)(\bar{bpy})(CO)H + CO + CO_3^{2-} \qquad (56)$$

Digital simulation techniques allow the conclusion
that the cyclic voltammetry data is consistent with
mechanism that is first order in CO_2 and first order in
the di-reduced complex $[Os(bpy)_2(CO)H]^-$. Beyond the rate
determining step, however, two plausible mechanistic
steps can occur, either unimolecular in Os (Eq. 55), or
bimolecular (Eq. 56). An unprecedented feature of the

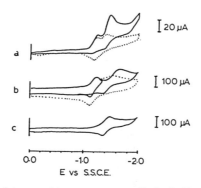

FIGURE 5. Cyclic voltammogram of 0.5mM [**fac**-Re(bpy)(CO)₃(CH₃CN)]⁺ taken in CH₃CN/0.1M TBAH with a Pt button working electrode.
  a.) Dotted line is single scan under argon taken at 0.2 V/sec; solid line is a CO₂ saturated solution.
  b.) Dotted line is single scan under argon taken at 10 V/sec; solid line is a CO₂ saturated solution.
  c.) CO₂ saturated solution after five sweeps at 10 V/sec.

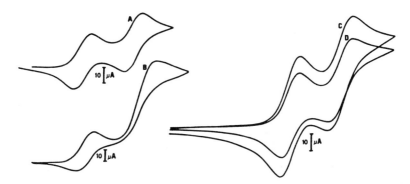

FIGURE 6. Cyclic voltammograms of ca. 4 mM [Os(bpy)₂(CO)H]⁺ taken in CH₃CN/0.1M TBAH with a Pt button working electrode. A and B show the complex under argon with 0.045 M CO₂ solution, respectively, at 0.1 V/sec. C and D show the effect of scan rate and CO₂ concentration, i.e., both were recorded in a 0.014 CO₂ solution but at scan rate of 0.2 V/sec and 0.1 V/sec, respectively.

mechanism is that CO_2 interacts in an <u>associative</u> manner
with the Os center (Eq. 54). Furthermore, labelling
studies have demonstrated that the metal-hydride linkage
is <u>maintained</u> during the catalysis, i.e., insertion of
CO_2 into the bond does not occur. This strongly suggests
that the intermediate $[Os(bpy)_2(CO)(CO_2)H]^-$ has a
non-hydridic Os-H bond which could be the consequence of
a <u>strongly electron withdrawing CO_2 ligand</u>.

Preparative scale electrolyses using a Pt gauze
electrode in a $CH_3CN/TBAH$ medium at potentials from -1.4
to $-1.6V$ gives activities of up to 0.4 turnovers/min with
current yields for CO of ca. 90%. We have not tested the
long term catalyst stability but in our short term
mechanistic work up to 20 turnovers results in no
catalyst decomposition and >95% recovery of the starting
complex.

In the presence of H_2O as a proton donor, kinetic
and product evidence is consistent with a formate
producing pathway that functions in competition with CO
formation. The proposed mechanism is shown below in Eqs.
57 and 58 where the k_3 step is rate limiting.

$$[Os(bpy)_2(CO)(CO_2)H]^- + H_2O \underset{k_{-2}}{\overset{k_2}{\rightleftharpoons}} I_a + {}^-OH \qquad (57)$$

$$I_a + e^- \xrightarrow{k_3} [Os(bpy)_2(CO)H] + {}^-O-\overset{\overset{\displaystyle O}{\|}}{C}-H \qquad (58)$$

Bulk electrolysis experiments with added H_2O give
formate with a Faradaic efficiency of up to ca. 25%. In
our current efforts, which are focussed on understanding
the kinetic branch which produces formate at the expense
of CO, several points are already apparent, these
include: 1.) the oxalate pathway which is rapid for the
dimerization of CO_2^- which has been electrochemically
generated (<u>6d</u>) is nearly totally suppressed at the
expense of CO and formate production, and, 2.) the use of
H_2O as a proton source implies a C-protonation route
followed by electron transfer.

<u>MODIFIED ELECTRODES BASED ON POLYMERIC ELECTROCATALYSTS</u>.

Electropolymerization of fac-Re(vbpy)$(CO)_3$Cl (vbpy is
4-methyl, 4'-vinyl-2, 2'-bipyridine) on Pt or glassy
carbon surfaces yields a chemically modified electrode

with a high specific activity for CO production (3a).   A
typical electrode response for a modified surface is
shown in Fig. 7.   Intitial rates obtained by bulk
electrolytic experiments at -1.5V give 20 turnovers/min,
but after ca. 350 turnovers the system deactivates.
    Copolymerization of the Re complex with, for
example, the catalytically inactive complex
$[Ru(bpy)_2(vpy)_2]^{2+}$ (vpy is 4-vinyl-pyridine), can yield
modified electrodes ratios of Re to Ru of up to 1:3.   In
these systems considerable improvement both in rates and
in stability can be achieved, as exemplified by up to ca.
$4x10^3$ turnovers at an initial rate of 100-200
turnovers/min.   Of particular interest is that even when
the catalytic activity of the system is lost the metal
complex is still on the electrode surface, apparently in
an inactive form.   We have also produced up to ca. 6%
oxalate during the catalytic reaction, and it is tempting
to speculate that the proximity effect of two bound CO_2
molecules is to enhance the C-C coupling reaction.
Recently Deronzier and coworkers (3p) have obtained
results similar to ours using oxidatively polymerized
pyrrole substituted bipyridine complexes.
    Another film-based electrocatalyst which we have
investigated involves oxidatively electropolymerized
Ni(TAP) (where TAP is
tetrakis-(*o*-aminophenyl)-tetraphenylporphine) on Pt
electrode surfaces (13).   For this modified electrode the
major product is formate (50% current yield), while minor
products include H_2 (35%) and CO (2%).   The conditions
for the reduction include using CH_3CN as the solvent *with
added water as a proton source.*   By electrolyzing at a
potential of -1.35V initial rates of 200-300
turnovers/min can be achieved with high stability.
    Both the Re-vbpy and the Ni(TAP) electrodes have
several orders of magnitude greater stability and
activity than their solution analogs and are within an
order of magnitude of the CO producing Co-phthalocyanine
modified electrode reported by Lieber and Lewis (3n).
Efforts are underway to maximize the performance of these
novel film-based electrodes.

DESIGN OF FUTURE CO_2 REDUCTION CATALYSTS.

Our mechanistic work has resulted in several insights
which may be of value in the design of future catalyst
systems for the reduction of CO_2 past the formate or CO
stage.   They include the following points: 1.) the use of
"electron reservoir" complexes acting as catalysts in
which more than one electron is held on ancillary
ligands, the central metal atom, or both, 2.) the fact

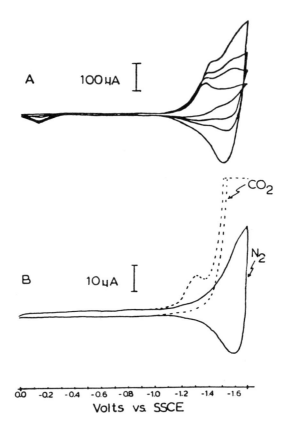

FIGURE 7. Cyclic voltammograms of chemically modified electrodes prepared by electropolymerization of fac-Re(vbpy)(CO)$_3$Cl in CH$_3$CN/0.1M TBAH with a Pt button working electrode.
- a.) Formation of the surface layer during single scanning.
- b.) The catalytic current (dashed line) observed in a CO$_2$ saturated CH$_3$CN solution.

that complex stability can be maximized by using chelate
type ligands, or monodentate ligands (e.g. CO) that have
exceptionally strong bond energies, 3.) associative
mechanisms which involve bond making and breaking at one
site in the molecule may minimize deactivation routes,
4.) bimolecular one-electron steps to make CO are
facilitated by metal complexes at the expense of oxalate
formation, 5.) formate can be electrocatalytically
produced either by direct insertion mechanisms, or by CO_2
coordination followed by electron-protonation steps,
although in the latter case the subsequent mechanistic
details still remain obscure, 6.) for further reduction
of CO_2 to methanol, catalytic intermediates which
currently produce formate are likely candidates. 7.)
polymeric systems offer the promise of high turnover
numbers and added stability toward degradative pathways.
    Complexes which contain electrons which are
localized at chemical sites possessing redox potentials
necessary for CO_2 reduction (electron reservoirs) can
clearly act as catalysts for CO_2 reduction. At this
early stage, however, a distinction between the
effectiveness of different electron reservoir complexes
can not be made strictly on the basis of the number or
location of the reducing equivalents within the molecule,
rather such choices depend on the availabilty of a
coordination site for the CO_2 ligand and the subsequent
mechanistic paths that form products, or, that result in
catalyst deactivation.
    We have shown in several cases, e.g. that of
$Os(bpy)_2(CO)H^+$ and $\underline{fac}$-Re(bpy)(CO)$_3$H, that associative
mechanisms result in no detectable degradative pathways.
This indicates that associative mechanisms which have a
high degree of specificity for a single coordination (and
hence reaction) site within the molecule should be future
targets in the design of highly stable catalysts. The
choice of second and third row transition metals is a
logical one here, since degradative exchange of the
ancillary ligands with solvent or with other potential
ligands, like CO, can be minimized compared with their
lighter congeners.
    It may be that complexes like the Os and Re examples
cited here which produce formate are good candidates for
further reduction chemistry that can occur in formal
two-electron steps. In a reactivity sense the analogies
shown below in structures a1-a8 suggest the possible
existence of related pathways worth pursuing.

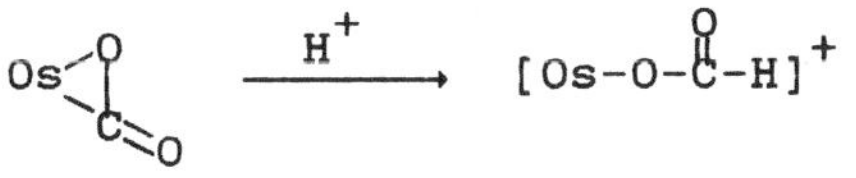

(a1)

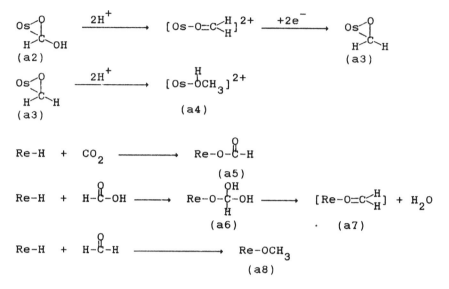

ARTIFICIAL PHOTOSYNTHESIS. A primary, long term goal of electrocatalytic CO_2 reduction studies is the development of homogeneous or heterogeneous chemical systems which convert light energy, carbon dioxide, and other common materials such as water into fuel and chemical precursors. The future advances in this area will likely be in the understanding of CO_2 reduction mechanisms, and the development of photosensitizer systems capable of driving the appropriate electrocatalysts.

ACKNOWLEDGMENTS. The authors wish to thank the Gas Research Institute for generous funding.

REFERENCES AND NOTES

(1) *Biological Carbon Dioxide Reduction*:
    (a) Thauer, R.K.; Diekert, G.; Schoenheit, P. Trends Biochem. Sci. 1980, 5, 304-306.
    (b) Romesser, J.A.; Wolfe, R.S. J. Bacteriol. 1982, 152, 840-847.
    (c) Romesser, J.A.; Wolfe, R.S. J. Bacteriol. 1985, 161, 696-701.

(2) *General Reviews*:
    (a) Darensbourg, D.J.; Kudaroski, R.A. Adv. Organometal. Chem. 1983, 22, 129-168.
    (b) Floriani, C. Pure & Appl. Chem. 1983, 55, 1-10.
    (c) Palmer, D.A.; Van Eldik, R. Chem. Rev. 1983, 83, 651-731.
    (d) Ibers, J.A. Chem. Soc. Rev. 1982, 57-73.

(e) Eisenberg, R.; Hendriksen, D.E. Adv. Cat. 1979, 28, 79-186.
(f) Volpin, M.E.; Kolomnikov, I.S. Organometallic Reactions 1975, 5, 313.
(g) Ziessel, R. Nouv. J. Chim. 1983, 7, 613-633.
(h) "Organic and Bio-Organic Chemistry of Carbon Dioxide", Edited by S. Inoue and N. Yamazaki, Halsted Press/John Wiley and Sons, New York. 1982, 280pp.
(i) Sneeden, R.P.A. in "Comprehensive Organometallic Chemistry", Edited by G. Wilkinson, Oxford Press, 1982, Vol. 8, Chapter 50.4.

(3) *Electrocatalytic Carbon Dioxide Reduction*:
(a) O'Toole, T.R.; Margerum, L.D.; Westmoreland, T.D.; Vining, W.J.; Murray, R.W.; Meyer, T.J. J. Chem. Soc., Chem. Commun. 1985, 865, 1416-1417.
(b) Sullivan, B.P.; Bolinger, C.M.; Conrad, D.; Vining, W.J.; Meyer, T.J. J. Chem. Soc., Chem. Commun. 1985, 1414-1416.
(c) Bolinger, C.M.; Sullivan, B.P.; Conrad, D.; Gilbert, J.A.; Story, N.; Meyer, T.J. J. Chem. Soc., Chem. Commun. 1985, 796-797.
(d) Ishida, H.; Tanaka, K.; Tanaka, T. Chem. Lett. 1985, 405-406.
(e) Beley, M.; Collin, J.-P.; Ruppert, R.; Sauvage, J.-P. J. Chem. Soc., Chem Commun. 1984, 1315-1316.
(f) Takahashi, K.; Hiratsuka, K.; Sasaki, H.; Toshima, S. Chem. Lett. 1979, 305-308.
(g) Wagenknecht, J.H.; Slater, S. J. Am. Chem. Soc. 1984, 106, 5367-5368.
(h) Eisenberg, R.; Fisher, B. J. Am. Chem. Soc. 1980, 102, 7363-7365.
(i) Meshitsuka, S.; Ichikawa, M.; Tamaru, K. J. Chem. Soc., Chem. Commun. 1974, 158-159.
(j) Tezuka, M.; Yajima, T.; Tsuchiya, A. J. Am. Chem. Soc. 1982,.104, 6834-6836
(k) Kapusta, S.; Hackerman, N. J. Electrochem. Soc. 1984, 131, 1511-1514.
(l) Becker, J.Y; Vainas, B.; Eger, R.; Kaufman, L. J. Chem. Soc., Chem. Commun. 1985, 1471-1472.
(m) Hawecker,J.; Lehn, J.; Ziessel, R. J. Chem. Soc., Chem. Commun. 1984, 328-330.
(n) Lieber, C.M.; Lewis, N.S. J. Am. Chem. Soc. 1984, 106, 5033-5034.
(o) Hiratsuka, K.; Takahashi, K.; Sasaki, H.; Toshima, S. Chem. Lett. 1977, 1137-1140.
(p) Cosnier, S.; Deronzier, A.; Moutet, J.-C. J. Electroanal. Chem. 1986, 207, 315.
(q) Andre, J.-F.; Wrighton, M.S. Inorg. Chem. 1985, 24, 4288-4292.

(r) Ogura, K.; Yoshida, I. J. Mol. Cat. 1986, 34, 67-72.

(s) Surridge, N.A.; Meyer, T.J. Anal. Chem. 1986, 58, 1576-1578.

(t) Breikss, A.I.; Abruna, H.D. J. Electroanal. Chem. 1986, 201, 347-359.

(u) Pearce, D. J.; Pletcher, D. J. Electroanal. Chem. 1986, 201, 317-330.

(v) Kusuda, K.; Ishihara, R.; Yamaguchi, H. Electrochimica Acta 1986, 31, 657-663.

(w) Parkinson, B.A.; Weaver, P.F. Nature 1984, 309, 148.

(4) See, for example:
(a) Gilbert, J.A.; Eggleston, D.S.; Murphy, Jr., W.R.; Geselowitz, D.A.; Gersten, S.W.; Hodgson, D.J.; Meyer, T.J. J. Am. Chem. Soc. 1985, 107, 3855-3864.

(b) Murphy, Jr., Takeuchi, K.J.; Meyer, T.J. J. Am. Chem. Soc. 1982, 104, 5817-5819.

(5) Latimer, W.M., "Oxidation Potentials", Prentice-Hall: Edgewood Cliffs, N.J., 1952.

(6) Electrochemical Reduction of Carbon Dioxide:
(a) Eggins, B. J Electroanal. Chem. 1983, 24, 1190-1193.

(b) Russel, P.G.; Kovac, N.; Srinivasan, S.; Steinberg, M. J. Electrochem. Soc. 1977, 124, 1329-1338.

(c) Jordan, J.; Smith, P.T. Proc. Chem. Soc. 1960, 246.

(d) Amatore, C.; Saveant, J.-M. J. Am. Chem. Soc. 1981, 103, 5021-5023.

(e) Goodridge, F.; Presland, G. J. Applied Electrochem. 1984, 14, 791-796.

(f) Kaiser, U. Von; Heitz, H. Ber. Bunsen. Gesell., 1973, 77, 818-823.

(g) Lamy, E.; Nadjo, L.; Saveant, J.M. J. Electroanal. Chem. 1977, 78, 403-407.

(h) Ryu, J.; Andersen, T.N.; Eyring, H. J. Phys. Chem. 1972, 76, 3278-3286.

(i) Bewick, A.; Greener, G.P. Tet. Lett. 1969, 53, 4623-4626.

(j) Ratzenhofer, M.; Kisch, H. Angew. Chem. Int. Ed.,Eng. 1980, 19, 317.

(k) Van Rysselberghe, P.; Alkire, G.J. J. Am. Chem. Soc. 1944, 66, 1801.

(l) Teeter, T.E.; Van Rysselberghe, P. J. Chem. Phys. 1954, 12, 759.

(m) Dehn, H.; Gutmann, V.; Kirch, H.; Schober, G. Monatsh. Chem. 1962, 93, 1348.

(n) Thonstad, J. J. Electrochem. Soc. 1964, 111, 955.

(o) Haynes, L.V.; Sawyer, D.T. Anal. Chem. 1967, 39, 332-338.

(p) Summers, D.P.; Leach, S.; Frese, Jr., K.W. J. Electroanal. Chem. 1986, 205, 219-232.

(q) Canfield, D.; Frese, Jr., K.W. J. Electrochem. Soc. 1983, 130, 1772.

(r) Canfield, D.; Frese, Jr., K.W. J. Electrochem. Soc. 1984, 131, 2518.

(s) Paik, W.; Andersen, N.; Eyring, H. Electrochim. Acta 1969, 19, 1217.

(t) Hori, Y.; Kikuchi, K.; Murata, A.; Suzuki, S. Chem. Lett. 1986, 897.

(u) Frese, Jr., K.W.; Leach, S. J. Electrochem. Soc. 1985, 132, 259.

(v) Gressin, J.C.; Michelet, D.; Nadjo, L.; Saveant, J.M. Nouv. J. Chim. 1979, 3, 545.

(7) *Carbon Dioxide Complexes*:

(a) Bianchini, C.; Meli, A. J. Am. Chem. Soc. 2698-2699, 1984, 106.

(b) Beck, W,; Raab, K.; Nagel, U.; Steimann, M. Angew. Chem. Int. Ed., Engl. 1982, 21, 526-527.

(c) Aresta, M.; Nobile, C.F. Inorg. Chim. Acta 1977, 24, L49-L50.

(d) Chatt, J.; Hussain, W.; Leigh, J. Transition Met. Chem. 1983, 8, 383-384.

(e) Ashuri, S.; Miller, D.J. Inorg. Chim. Acta 1984, 88, L1-L3.

(f) Albano, P.; Manassero, M. Inorg. Chem. 1980, 19, 1069-1072.

(g) Mahler, J.M.; Lee, G.R.; Cooper, N.J. J. Am. Chem. Soc. 1982, 104, 6797.

(h) Lee, G.R.; Cooper, N.G. Organometallics 1985, 4, 794.

(i) Bodner, T.; Coman, E.; Menard, K.; Cutler, A. Inorg. Chem. 1982, 21, 1275.

(j) Green, M.L.H.; Mackenzie, R.E.; Poland, J.S. J. Chem. Soc., Chem. Comm. 1976, 1993.

(k) Grice, N.; Kao, S.C.; Pettit, R. J. Am. Chem. Soc. 1979, 101, 1627.

(l) Calabrese, J.C.; Herskovitz, T.; Kinney, J.B. J. Am. Chem. Soc. 1983,.

(m) Fachinetti, G.; Floriani, C.; Zanazzi, P.F.; Zanzari, A.R. Inorg. Chem. 1979, 18, 3469-3475.

(n) Alvarez, R.; Carmona, E.; Poveda, M.L.; Sanchez-Delgado, R. J. Am. Chem. Soc. 1984, 106, 2731-2732.

(o) Gambarotta, S.; Arena, F.; Floriani, C.; Zanazzi, P.F. J. Am. Chem. Soc. 1982, 104, 5082-5092.

(p) Bristow, G.S.; Hitchcock, P.B.; Lappert, M.F. J. Chem. Soc., Chem. Commun. 1979, 1145-1146.

(q) Alvarez, R.; Carmona, E.; Gutierrez-Puebla, E.; Marin, J.M.; Monge, A.; Manuel, L.P. J. Chem. Soc., Chem. Commun. 1984, 924, 1326-1327.

(r) Aresta, M.; Nobile, C.F.; Albano, V.G.; Forni, E.; Manassero, M. J. Chem. Soc., Chem. Comm. 1975, 636.

(s) Aresta, M.; Nobile, C.F. *J. Chem. Soc., Dalton Trans*. 1977, 708.
(t) Fachinetti, G.; Floriani, C. *J. Am. Chem. Soc*. 1978, 100, 7405.
(u) English, A.D.; Herskovitz, T. *J. Am. Chem. Soc*. 1977, 99, 1648.
(v) Gambarotta, S.; Floriani, C.; Chiesi-Villa, A.; Guastini, C. *J. Am. Chem. Soc*. 1985, 107, 2985.

(8) *General Carbon Dioxide Complex Chemistry*:
(a) Maher, J.M.; Cooper, N.J. *J. Am. Chem. Soc*. 1980, 102, 7606.
(b) Yamamoto, A.; Tazuma, S.K.; Pu, L.S.; Ikeda, S. *J. Am. Chem. Soc*. 1971, 93, 371.
(c) Evans, G.O.; Walter, W.F.; Mills, D.R.; Streit, C.A. *J. Organomet. Chem*. 1978, 144, C34.
(d) Nicholas, K.M. *J. Organomet. Chem*. 1980, 188, C10.
(e) Chang, B.-H. *J. Organomet. Chem*. 1985, 291, C31-33.
(f) Keene, F.R.; Cruetz, C.; Sutin, N. *Coord. Chem. Rev*. 1985, 64, 247.

(9) Gambarrotta, S.; Arena, F.; Gaetani-Manfredotti, A. *J. Chem. Soc.,Chem. Commun*. 1982, 835.

(10)*Carbon Dioxide Insertion Reactions*:
(a) Lyons, D.; Wilkinson, G. *J. Chem. Soc., Dalton Trans*. 1985, 587-590
(b) La Monica, G.; Ceinini, S.; Porta, F.; Pizzotti M. *J. Chem. Soc., Dalton Trans*. 1976, 1777-1782.
(c) Merrifield, J.H.; Gladysz, J.A. *Organometallics* 1983, 2, 782-784.
(d) Behr, A.; Kanne, U.; Thelen, G. *J. Organomet. Chem*. 1984, 269, C
(e) Willis, W.; Nicholas, K.M. *Inorg. Chim. Acta* 1984, 90, L51-L53.
(f) Sullivan, B.P.; Silliman, S.; Thorp, H.; Meyer, T.J. submitted.
(g) Darensbourg, D.J.; Kudarowski, R. *J. Am. Chem. Soc*. 1984, 106, 3672-3673.
(h) Komiya, S.; Yamamoto, A. *J. Organomet. Chem*. 1972, 46, C58-C60.
(i) Kolomnikov, I.S.; Gusev, A.I.; Aleksandrov, G.G.; Lobeeva, T.S.; Struchkov, Y.T.; Vol'pin, M.E. *J. Organomet. Chem*. 1973, 59, 349-351.
(j) La Monica, G.; Ardizzoia, G.A.; Cariati, F.; Cenini, S.; Pizzoti, M. *Inorg. Chem* 1985, 24, 3920-3923.
(k) Gambarotta, S.; Strologo, S.; Floriani, C.; Chiesi-Villa, A.; Guastini, C. *J. Am. Chem. Soc*. 1985, 107, 6278-6282.

(l) Darensbourg, D.J.; Hanckel, R.K; Bauch, C.G.;
    Pala, M.; Simmons, G.; White, G.N. J. Am. Chem.
    Soc. 1985, 107, 7463-7472.
(m) Bochkarev, M.N.; Fedorova, E.A.; Radkov, Yu.F.;
    Khorshev, S.Ya.; Kalinina, G.S.; Razuvaev, G.A. J.
    Organomet. Chem. 1983, 258, 29-33.
(n) Darensbourg, L.; Prengel, C. Organometallics 1984,
    3, 934-936.
(o) Gaus, P.L.; Kao, S.C.; Youngdahi, K.; Darensbourg,
    M.Y. J. Am. Chem. Soc. 1985, 107, 2428-2434.
(p) Kao, S.C.; Gaus, P.G.; Youngdahi, K.;
    Darensbourg, M.Y. Organometallics 1984, 3,
    1601-1603.
(q) Sullivan, B.P.; Meyer, T.J. J. Chem. Soc., Chem.
    Commun. 1984, 1244-1245.
(r) Darensbourg, D.J.; Grotsch, G. J. Am. Chem. Soc.
    1985, 107, 7473-7476.

(11) Kew, G.; DeArmond, K.; Hanck, K. J. Phys. Chem.
    1974, 78, 727.

(12) *Photochemical Reduction of Carbon Dioxide*:
    (a) Hawecker, J.; Lehn, J.-M.; Ziessel, R. J. Chem.
        Soc., Chem.Commun. 1985, 56-58.
    (b) Sullivan, B. P.; Meyer, T. J. J. Chem. Soc., Chem.
        Commun. 1984, 1244-1245.
    (c) Bradley, M.G.; Roberts, D.A.; Geoffroy, G.L. J.
        Am. Chem. Soc. 1981, 103, 379-384.
    (d) Hawecker, J.; Lehn, J.; Ziessel, R. J. Chem. Soc.,
        Chem. Commun. 1983, 536-538.
    (e) Lehn, J.; Ziessel, R. Proc. Natl. Acad. Sci., USA
        1982, 79, 701-704.
    (f) Akermark, B.; Eklund-Westlin, U.; Baeckstrom, P.;
        Lof, R. Acta. Chemica Scand. 1980, 34, 27-30.
    (g) Kutal, C.; Weber, M.A.; Ferraudi, G.; Geiger, D.
        Organometallics 1985, 4, 2161-2166.
    (i) Hukkanen, H; Pakkanen, T.T. Inorganica Chimica
        Acta 1986, 114, L43-L45.

(13) Sullivan, B.P., unpublished data.

(14) *Photoelectrochemical Reduction of Carbon Dioxide*:
    (a) Bradley, M.G.; Tysak, T.; Graves, D.J.;
        Vlachopoulos, N.A. J. Chem. Soc., Chem. Commun.
        1983, 349-350.
    (b) Inque, T.; Fujishima, A.; Konishi, S.; Honda, K.
        Nature 1979, 277, 637-638.
    (c) Aurian-Blajeni, B.; Habib, M.A.; Taniguchi, I.;
        Bockris, J.O'M. J. Electroanal. Chem. 1983, 157,
        399-404.
    (d) Taniguchi, I.; Aurian-Blajeni, B.; Bockris, J.O'M.
        J. Electroanal. Chem. 1983, 157, 179-182.
    (e) Taniguchi, Y.; Yoneyama, H.; Tamura, H. Bull.
        Chem. Soc., Jpn. 1982, 55, 2034-2039.

(f) Halmann, M. Nature 1978, 275, 115-116.
(g) Aurian-Blajeni, B.; Halmann, M.; Manassen, J. Solar Energy 1980, 25, 165-170.
(h) Zafrir, M.; Ulman, M.; Zuckerman, Y.; Halmann, M. J. Electroanal. Chem. 1983, 159, 373-389.
(i) Giner, J. Electrochimica Acta. 1963, 8, 857-865.
(j) Inoue, T.; Fujishima, A.; Honda, K. Nature 1979, 277, 637.
(k) Cabrera, C.R.; Abruna, H.D. J. Electroanal. Chem. 1986, 209, 101-107.

RECEIVED May 13, 1987

Chapter 7

Enzymatic Activation of Carbon Dioxide

Leland C. Allen

Department of Chemistry, Princeton University, Princeton, NJ 08544

The kinetics and electronic mechanisms of conventional chemical catalysts are contrasted with
those in enzymes. The analogy between certain
attributes of surfactants and phase-transfer
catalysis and enzyme active sites are made and the
limitations of surface catalysts and zeolites are
pointed out. The principle features that give
enzymes their unusual rate enhancements and
remarkable specificity are discussed and ways in
which these can be realized in man-made catalysts are
proposed. The catalytic activation of CO_2 by both
enzymatic and non-enzymatic means, including a
detailed analysis of the electronic reaction sequence
for the metalloenzyme carbonic anhydrase, is used to
illustrate the above themes.

In this article we discuss the unique features of enzyme catalysis
of CO_2 hydrolysis and how this may relate to the design of new
man-made CO_2 catalysts for various reactions of practical use. The
other papers in this volume are concerned with CO_2 catalysis at
metal and metal oxide surfaces using first, second, and third row
transition metals. Other catalytic processes treated in this
volume are electrochemical and photoelectrochemical reduction of
CO_2. It is hoped that further developments of these man-made
systems will benefit from a detailed analysis of the biological
catalyst carbonic anhydrase. This enzyme is a very old one that
has evolved to near perfection in its metabolic role of aiding the
solvation of CO_2 into blood. It is one of the simplest enzymes and
has one of the highest known turnover rates. We contrast the
hydrolysis of CO_2 by carbonic anhydrase with that by simple bases
to understand the factors contributing to the enormous rate
enhancements realized by enzymes, and we describe surfactant and
phase-transfer catalysis to show how these partially accomplish
some of the benefits of enzymes. We then give the details of the

0097–6156/88/0363–0091$06.00/0

recently discovered electronic-level reaction sequence employed by
carbonic anhydrase to convert CO_2 to $HCO_3^-+H^+$. Although this
specific chemical reaction is of little current commercial
interest, it illustrates several general principles for efficient
catalysis. A central feature of carbonic anhydrase and other
enzyme systems is their inherent three-dimensionality. Proposals
for utilizing this and other carbonic anhydrase features in the
synthesis of new catalysts are discussed.

Enzymatic Versus Non-enzymatic Catalysis of CO_2

We consider the well-known simple kinetic model for bimolecular
reactions between a catalyst, C, its substrate, S, and product, P:

$$C + S \underset{k_{-1}}{\overset{k_1}{\rightleftarrows}} CS \xrightarrow{k_{cat}} C + P$$

where CS is the catalyst-substrate complex, an entity that plays a
significant role in differentiating conventional and enzyme
catalysis. Applying the steady state condition for the concentration of CS leads to the rate expression:

$$v = \frac{k_{cat}[Co][S]}{K_d + [S]}$$

where $[C_o]$ is the specified initial catalyst concentration, $[S]$ the
concentration, and $K_d \equiv \dfrac{[C][S]}{[CS]} = \dfrac{k_{-1}+k_{cat}}{k_1}$, is the dissociation
equilibrium constant for the CS complex. The equation for v above
describes a rate versus $[S]$ behavior that rises linearly at small
$[S]$ and then bends over at high $[S]$ to a constant value which is
zero order in $[S]$. In most cases of interest $k_{cat} \gg k_{-1}$ and
$k_1 \gg k_{-1}$, thus $K_d = k_{cat}/k_1$. (We note that for enzymes this simple
model is the Michaelis-Menten result with $K_d = K_M$, the Michaelis
constant, and CS the Michaelis complex). At low substrate
concentrations the rate expression reduces to:
$$v = (k_{cat}/K_d)[C_o][S]$$
where $k_{cat}/K_d = k$, the second order rate constant. For ordinary
solution phase catalysis, the CS association constants, $1/K_d$, are
typically $10^{-1}-10^2 \approx 1$ (the latter value applies to oppositely
charged ions and is not relevant to CO_2 as substrate). In carbonic
anhydrase $K_M = 8 \times 10^{-3}$ M and $k_{cat} = 6 \times 10^5 s^{-1}$, therefore, $k = 7.5 \times 10^7$
$M^{-1}s^{-1}$. Note that the CS association constant is yielding a rate
enhancement of $\approx 10^2$. More typically this factor is 10^3-10^4. This
is the famous lock & key non-covalent binding (London dispersion
forces and hydrogen bonds) which gives rise to an enzyme's very
high substrate specificity and which holds the substrate in proper
alignment for the subsequent electronic rearrangement of its
bonds. It is one important manifestation of the inherent three
dimensionality of enzyme catalysis. Substitution of $K_d = k_{cat}/k_1$

into the rate equation at low substrate concentration gives the
second order rate constant as $k = k_1$, thus showing that for this
condition the formation of the CS complex is rate determining.
    Second order rate constants for several bases reacting with
CO_2 compared to carbonic anhydrase are shown in Table I.   The

Table I.   Second Order Rate Constants for CO_2
Reactions in H_2O at 25°C

Base	$k(M^{-1}s^{-1})$
OH^-[a]	8.5×10^3
CH_3NH_2[b]	2.0×10^4
$C_5H_5NH_2$[b]	5.3×10
$Co(NH_3)_5OH^+$[a]	2.2×10^2
H_2O[c]	6.7×10^{-4}
Carbonic anhydrase[d]	7.5×10^7

[a]R. B. Martin, J. Inorg. Nucl. Chem. **38**, 511 (1976).   [b]M. B. Jensen,
Acta Chem. Scand. **13**, 289 (1959).   [c]R. G. Khalifah, J. Biol. Chem.
246, 2561 (1971).   [d]R. G. Khalifah, Proc. Nat. Acad. Sci., US, **70**,
1986 (1973).

common feature in all of these is that the rate determining step
(except for the enzyme) is attack of the base on the carbon of CO_2
and we can see that the enzyme has a rate constant at least three
orders of magnitude greater than any of them.   The bases OH^- and
$Co(NH_3)_5OH^+$ (and $Zn(OH_2)_5OH^+$ which is closely analogous), as well
as carbonic anhydrase, lead to the hydrolysis of CO_2 and so further
comparisons can be made.   One such comparison is pK_a values and for
the three bases and enzyme these are: 15.7, 6.6, 6.9, and 7.2
respectively.   If we are interested in operation at physiological
pH, OH^- can not exist, but the cobalt and zinc complexes can and
they therefore (see paragraph below) can offer additional insight
into the unique features of enzymatic catalysis.
    A simple reaction sequence schematic for the uncatalyzed
hydrolysis (H_2O as base in Table I.) is:

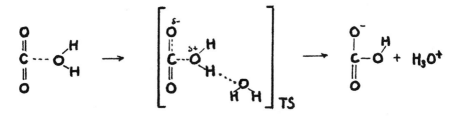

Note that the rate enhancement for carbonic anhydrase over the
uncatalyzed reaction is 1.12×10^{11}, a not atypical value for many
enzymes.   For the zinc-water-hydroxyl complex a reaction sequence
schematic is:

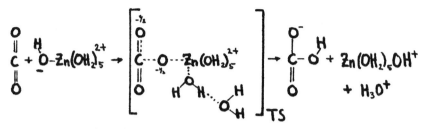

Not unexpectedly, the carbonic anhydrase reaction sequence is more
complicated. We give it and make comparison with the above sequence
in the next section.

The rate constant k_{cat} is that identified with the _electronic_
rearrangement undergone in the reaction. From $k = k_{cat}/K_d$ with
$K_d = 1$ the k values of Table I translate numerically into $k_{cat}(s^{-1})$
and compared to carbonic anhydrase ($k_{cat} = 6 \times 10^5 s^{-1}$) we see that
enhancements range from 30 to 10^4. In addition to the
non-covalent, lock-and-key, enzyme-substrate fitting that
constitutes the Michaelis-Menton CS complex, there are other
structural features of the enzyme-substrate complex which lead to
high k_{cat}. Thus, an enzyme's active site cavity (for carbonic
anhydrase, a cone-shaped indentation 12 Å in diameter and 12 Å deep
with a zinc atom at the bottom) is lined with acids and bases which
induce the electronic rearrangements that produce the catalysis.
These are positioned three dimensionally such that they have their
maximum catalytic effect when the substrate has electronically
deformed to its transition state conformation. (For carbonic
anhydrase the amino acid side chains which carry out the acid-base
catalysis are His 64, Glu 106(-), Thr 199, a hydroxyl attached to
zinc (a water molecule depronated by the zinc ion) and a water
molecule hydrogen bonded to Thr 199).

With the above insight into the nature of k_{cat}, we can gain
further knowledge of enzyme action by considering the rate
expression at high substrate concentration. For high [S] all the
catalytic sites become filled with substrate and the rate is solely
determined by the _turnover rate_, k_{cat}. In this case:

$$v = k_{cat} [C_o]$$

We now refer back to the low [S] expression where we showed the
rate enhancement advantage of having the Michaelis CS complex and
we also noted that $k = k_1$, the rate of complex formation. At low
[S], increasing the Michaelis complex association constant
increases the rate linearily, but because of catalytic active site
saturation this is not true for larger [S]. If the CS binding is
too tight, saturation will occur at too low an [S]. The CS complex
must be just strong enough (but not stronger) than required to
accomplish its purpose of preventing an unnecessarily large number
of C+S collisions before reaction and of aligning the substrate for
its subsequent smooth _electronic_ movement to the transition state.
We may now imagine the overall enzymatic process by starting with
the usual graph of ΔH versus reaction coordinate which shows a
single barrier between the reactant and product. Ordinary
(non-enzymatic) catalysis is manifest by a simple lowering of this

single barrier. In enzymes, k_{cat} values reflect a greatly lowered
main barrier but there is also an additional small ΔH maximum and
the Michaelis complex ΔH binding minimum which is introduced
between the reactant side of the ΔH versus reaction coordinate
curve and the main (k_{cat}) transition state. The new Michaelis
complex binding minimum in ΔH compensates for the large decrease in
entropy that occurs on formation of the Michaelis complex. The new
small maximum in ΔH is the activation barrier to Michaelis complex
formation.

The various aspects of enzyme action discussed above may be
viewed as nature's attempt to create a chemical reaction pathway
somewhere between that of a gas phase reaction and one occurring in
pure water solution. The problem of numerous large barriers must
be overcome in the gas phase, while in water, charge separation is
so screened out that large barriers can also arise. In the enzyme
all of these barriers are made small and the reaction proceeds
along a relatively smooth middle path.

Surfactants and detergents are widely used commercially to
provide a favorable environment for numerous organic and inorganic
reactions that could otherwise not be carried out. Surfactants are
made up of hydrophilic and hydrophobic groups which form micelles
in water. The small volume and hydrophobic nature of the micelle
interior provides a region that typically enhances rates by 10^3. A
parallel event is happening in enzymes and is part of the effect we
have ascribed to Michaelis complex binding and entropy reduction.
The enzyme active site excludes almost all water molecules (except
those which are an integral part of the reaction mechanism or those
that act as structural elements in orienting substrate for chemical
reaction). Exclusion of water provides a dielectric constant of
near unity and allows easy separation of charges - the key feature
required to lower the energy for charge rearrangements in the
transition state. Phase-transfer catalysis is another much
employed industrial process that also bears some analogy to enzyme
active sites. Again, the advantage of carrying out ionic reactions
in a low dielectric is realized in the presence of two immissible
solvent media (generally water and a hydrocarbon).

The Electronic Reaction Mechanism of Carbonic Anhydrase

In spite of the fact that enzymology is an old and well-established
discipline, electronic-level understanding of enzyme mechanisms
(such as we have for a number of organic and inorganic reactions)
is only now just beginning to emerge for one or two enzymes.
Carbonic anhydrase is one of these. Likewise, although there are
numerous textbooks which put forth mechanistic hypotheses, none of
them offer long-standing, universally agreed-upon enzyme reaction
sequence schematics at the electronic level and, at present, one
must rely on the current literature.

For carbonic anhydrase, a great deal of effort has been
expended over many years in many laboratories, and the work on its
mechanism has recently been reviewed (1-2). The computational
techniques for constructing the reaction potential energy surfaces
and their relation to experimental measurements has also been
treated (2-4). For a long time there was a conflict between
spectroscopists - particularily those employing NMR - and

kineticists as to the role of water molecules in the reaction, but a systematic computational testing of the various mechanistic hypotheses appears to have led to a concensus among the experimentalists. We focus our attention here on that part of the reaction which converts CO_2 into HCO_3^- and isolate those electronic features which might be most easily simulated in man-made catalysts. (The other part of the mechanism involves buffer and the movement of a proton in and out of the active site and it is unique to carbonic anhydrase's physiological role in metabolism).

Figure 1 shows the arrangement of amino acid residues in the carbonic anhydrase active site and Figure 2 reduces this to a two-dimensional working schematic. Initially the active site contains a zinc-bound hydroxyl, and a water molecule bound distantly as a zinc fifth ligand. This water molecule is hydrogen bonded to Thr 199 which in turn is hydrogen bonded to Glu 106(-). Glu 106(-) is exposed to solvent, is relatively near zinc (2+) and is hydrogen bonded to the backbone nitrogen of Arg 246. All of these bonds stabilize the negative charge on Glu 106(-). The zinc-bound hydroxyl is hydrogen bonded through a water chain to His 64. In the absence of the CO_2 substrate an additional water molecule will be hydrogen bonded to the hydroxyl and this must be displaced by the incoming CO_2.

Figure 3 shows the detailed reaction sequence. Incoming CO_2 will be attacked by the zinc-bound hydroxyl (B) forming a cyclic intermediate with the fifth coordinate water. Additional hydrogen bonds with one or more water molecules in the active site link the bicarbonate proton to the exocarboxylate oxygen. As the cyclic intermediate is formed a unit of negative charge moves from the hydroxyl to bicarbonate. This will be compensated by a partial proton drift from the fifth coordinate water through Thr 199 to Glu 106. The proton affinity of the hydroxide will also drop because of C-O formation, thereby allowing a proton transfer to the exocarboxylate oxygen (C). His 64 stabilizes the transition state for this transfer. The loss of the bicarbonate product from the active site passes through the transition structure (D). As the zinc oxygen bond breaks, a negative charge (shown as $\approx \frac{1}{2}$) will develop on the oxygen. A proton from the zinc-bound water transfers through Thr 199 to Glu 106 regenerating the zinc-bound hydroxyl. This "around the corner S_N2 reaction" leads to a lowered binding energy for the bicarbonate product and thus to its release. In this process zinc merely trades anionic ligands rather than having to break a full ionic bond. The reaction cycle is completed in two steps: the addition of a water molecule and loss of a proton to solvent.

The reaction sequence shown in Figure 3 uses two water molecules as an intimate and rather subtle part of the mechanism and it is well to bring out how this occurs. The first water is the lower one in the diagram: it gets deprotonated by Zn^{2+} and the resulting hydroxyl acts as the principle attacking nucleophile in the reaction. This molecule is further deprotonated in the proton transfer to the exocarboxylate oxygen and therefore a strong Zn-O bond is established. This Zn-O bond must in turn be broken to release product. The electronic redistribution which accomplishes this is the `around the corner S_N2 reaction´ - one of nature's clever inventions that mankind will want to copy. The lower water

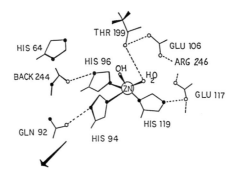

Figure 1.   Perspective view of carbonic anhydrase active
site.   The arrow points toward the opening of the
cavity.

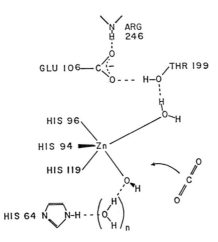

Figure 2.   Two dimensional schematic of carbonic anhydrase
active site and CO_2 substrate showing those parts
of the enzyme specifically involved in
catalysis.

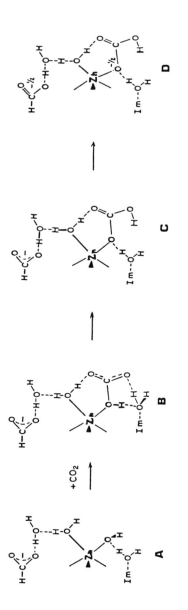

Figure 3.   Electronic reaction sequence for the catalytic
conversion of CO_2 to HCO_3^- by carbonic anhydrase.

also contains the `oxygen of hydrolysis´ - that oxygen which
becomes incorporated into the product - and as such isotopic
labeling readily identifies it. The second, upper, water acts as
part of a 'working medium': it makes possible a cyclic transition
state and its upper hydrogen moves off and on to its oxygen in
response to the relative proton affinity of the forming cyclic
transition state versus that of the Glu 106 - Thr 199 change
relay. Zn-O for this water starts out as a long bond, but becomes
a short bond as it passes through the transition state of (D).
When this bond becomes short it allows the lower Zn-O bond to
become long and break.

Catalytic Principles and Design of More Powerful Catalysts

It is now apparent why $Zn(OH_2)_5(OH)^+$ in water solution is a
relatively inefficient catalyst compared to carbonic anhydrase:
although the $O \cdots Zn$ bond in the transition state of the reaction
schematic given in the section above is shown as a dotted line,
there is no electronic rearrangement path available to aid in
breaking this bond (the TS could also have been written in a cyclic
form - Z_nOCOZ_n - but the same problem arises). The `around the
corner S_N2 reaction´ solves this problem. A closely related
feature is the practically constant charge of $\approx +1$ maintained on
zinc throughout the entire reaction sequence (which is another way
of saying that the activation barriers for the electronic
rearrangements in going from reactant to product are small). These
observations are generalized by the principle (5) of `two bond
maneuver´ (Figure 4) which exists in all enzymes we have studied.
If one bond is broken and one new bond made in a catalytic process
then it is necessary to supply enough energy to break that bond
even though a nearly equal amount is returned when another bond is
made. This can be avoided by simultaneously breaking and making
two bonds (one in the enzyme and one in the substrate) because each
atom maintains a roughly constant charge around itself. Moreover,
we can see from the atomic charge assignments in Figure 4 that bond
polarities are also maintained and this implies the opportunity for
low ΔH barriers.
     What are the essential chemical properties provided by zinc
and its ligands in carbonic anhydrase? (It may be noted that
replacement of zinc by cobalt and cadmium has been carried out and
for the cobalt substitution there is no change in properties while

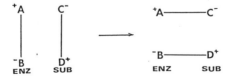

Figure 4. The 'Two Bond Maneuver' in enzymes (D. E. Green,
            Proc. Nat. Acad. Sci, U.S. 78, 5344 (1981))
            whereby activation barriers are made low by the
            simultaneous breaking and making of two bonds in
            the enzyme and substrate.

the cadmium enzyme still functions but at a low level). Zinc is acting as a strong Lewis acid (effective charge + 1) and in addition this atom needs to be of sufficient size and polarizability to maintain a roughly constant charge without large energy changes. This relatively simple functionality is the central role for zinc in all its enzymes and the same statement can be made for Mg and for about half of the enzymes utilizing Mn, Fe, Co, and Cu. Also because of these simple requirements, it is not surprising to find that nature has evolved two classes of enzymes that carry out the same chemical reaction by the same general mechanism, one of which utilizes zinc and the other NH_4^+ (in the form of a Schiff base). These are the fructose diphosphate aldolase enzymes in the glycolytic pathway which cleave a C-C bond and leave a C=O on one of the products.

Another catalytic principle can be learned from carbonic anhydrase and two other zinc enzymes which have been well-characterized: carboxypeptidase A and liver alcohol dehydrogenase. In each of these, Zn is held into its enzyme by three chemically inert ligands, but these ligands differ in their charge so that the formal charge on Zn in these three cases is 2+, 1+, and 0 respectively. Zinc is acting as a successively weaker Lewis acid, but the electronic rearrangements going on around zinc during catalysis have a remarkable similarity. In each case the zinc is 5-coordinate, 3 from the inactive ligands and 2 from two oxygen atoms that are actively participating in electron flows. One of the oxygen atoms is attached to the substrate and the other to a water molecule. (This is the upper water in the carbonic anhydrase mechanism which changes its Zn-O distance and its oxygen charge during the reaction). In effect, in each case, the catalyst is zinc plus the water molecule, with the fifth coordination site utilized by the substrate. The concept of a mobile satelite molecule containing an electronegative atom that acts as an electron switching agent may be applicable to other situations.

There are two other general themes of enzyme catalysis illustrated by carbonic anhydrase that bear directly on contemporary commercial catalysts. The first is the inherent three dimensionality that we have noted several times. Almost all of the many, many varieties of industrial catalysts utilize the properties of a two-dimensional surface, with the catalytic event typically taking place at the potential energy discontinuities created by `steps` or `islands` on the surface. Using only two dimensions is a very severe limitation because one cannot control the environment surrounding the atoms. One needs to use all three dimensions fully to do this. This does not mean that excellent catalysts can't be developed, but it does mean that their design will be governed by pure trial and error to a very large extent and suggests that we are far less advanced in our technological capabilities than we might be.

The second theme comes from examining the properties of zeolites. Here, the three dimensional nature of these catalysts has long been exploited by the petroleum industry, mostly because they function as molecular sieves. Zeolites, like all of the mineral silicates of which they are a special class, are constructed with highly ionic bonds and this introduces two severe limitations. First, the number of possible structures made out of ionic bonds is much less diverse than those possible from covalent bonding, and

second, the large positive and negative charges distributed in more
or less fixed positions throughout the zeolites is not conducive to
realization of the subtle charge redistributions and barrier
lowerings required for really flexible and versatile catalysts.

From the above it seems clear that in order to build fully
controllable three dimensional cavities lined with appropriate
functional groups it is probably necessary to follow nature's lead
and use an organic polymer made up of units with common end groups
to permit easy construction of a chain. However, the polypeptide
chains of proteins have their own problems: it should not be
necessary to generate α-helices and β-pleated sheets in order to
get stable and reasonably compact structures. Therefore, simpler
polymer building blocks, without the possibility of forming
hydrogen bonds, might be useful. Likewise, one probably doesn't
need all twenty of the amino acid side chains used in proteins and
the functional groups employed are not ideal for many of the kinds
of man-made molecules that are desired. Thus, a smaller and
simpler set might be used for attachment to the organic polymer
building-block units.

Experimentation with a polymer such as described above would
start by using a very few links to explore what types of
configurations could be realized. This exploration would be
greatly aided by the powerful molecular mechanics computer programs
that have been under intensive development during the last few
years. It is not at all unreasonable to expect that conformations
for at least the simple polymers can be predicted accurately. The
combination of modern computational methods along with the
continuing strong advances in polymer chemistry will match the
incredible accomplishment of nature's evolution more quickly than
could at first be imagined. These man-made analogies to enzymes
will play a key role in the nanotechnology revolution (the
development of molecular machines and devices at the 10 Å level)
predicted a few years ago by Profs. Richard P. Feynman and Freeman
Dyson. References to the articles by Feynman, Dyson and several
other seminal thinkers in this area are given in a recent popular
account by K. E. Drexler (6).

Acknowledgments

The author wishes to thank the Office of Naval Research and the
National Institute of Health for financial support of this
research.

Literature Cited

1. Cook, C. M.; Lee, R. H.; and Allen, L. C. Inter. J. Quant.
    Chem. Sym. 1983, 10, 263.
2. Cook, C. M; and Allen, L. C. Biology and Chemistry of Carbonic
    Anhydrase, Tashian R. E. and Hewett-Emmet, D., Eds., Ann. N.Y.
    Acad. Sci. 1984, 429, 84.
3. Cook, C. M.; Haydock, K.; Lee, R. H.; and Allen, L. C. J. Phys.
    Chem. 1984, 88, 4875.
4. Allen, L. C. Ann. N. Y. Acad. Sci. 1981, 367, 383.
5. Green, D. E. Proc. Nat. Acad. Sci. U.S. 1981, 78, 5344.
6. Drexler, K. E. Engines of Creation; Anchor Press/Double Day
    Ca., 1986.

RECEIVED December 1, 1986

Chapter 8

Adsorptive and Catalytic Properties of Carbon Monoxide and Carbon Dioxide Over Supported Metal Oxides

D. G. Rethwisch [1] and J. A. Dumesic [2]

[1]Department of Chemical and Materials Engineering, University of Iowa, Iowa City, IA 52242
[2]Department of Chemical Engineering, University of Wisconsin, Madison, WI 53706

The interactions of CO and CO_2 with iron oxide and zinc oxide supported on SiO_2, Al_2O_3, TiO_2, MgO, and ZnO were examined using infrared spectroscopy, Mössbauer spectroscopy and water-gas shift kinetics. Ferrous cations supported on SiO_2, Al_2O_3, and TiO_2 were stable against oxidation to Fe^{3+} by mixtures of CO and CO_2. This stability is thought to be caused by the formation of surface compounds between iron oxide and the various supports. The low activity of supported iron oxide relative to magnetite and the stability of Fe^{2+} on the supports suggests a change in the dominant reaction mechanism for water-gas shift from the regenerative mechanism over magnetite (Fe_3O_4), to the associative mechanism over the supported samples. Either CO or CO_2 may interact with surface hydroxyl species to form adsorbed formate and bicarbonate species. These surface species are proposed to be intermediates for the water-gas shift reaction. Catalytic activities of the supported samples decreased as the acidity of the support or the electronegativity of the support cations increased. It is proposed that CO does not interact readily with the hydroxyl species on acidic supports.

Studies in recent years indicate that supporting a catalyst on a high surface area carrier may influence its catalytic properties. Much of this work has centered around so-called "strong metal-support interactions" which are manifested, for example, by a marked decrease in the extent of hydrogen and carbon monoxide adsorption on titania-supported Group VIII metals when reduced at high temperatures (1-3). Due to structural similarities between oxide catalysts and typical supports, which are generally oxides, strong support interaction should also be observed for these catalysts (e.g., see papers in the recent monograph by Grasselli and Brazdil

0097-6156/88/0363-0102$06.25/0

(4)). Indeed, Lund and Dumesic (5-8) reported that the water-gas shift activity of an iron oxide catalyst is reduced by several orders of magnitude when supported on silica. Strong interactions between molybdena and alumina have been documented for the calcined states of hydrotreating catalysts (e.g., 9-11). Also, interaction is manifested in many mixed oxides by enhanced acidity, compared to the acidities of the pure component oxides (12-14).

In the present work, a series of supported iron oxide catalysts was characterized by Mössbauer spectroscopy and infrared spectroscopy. Also, the effects of the support on the catalytic activity for the water-gas shift reaction ($CO + H_2O <=> CO_2 + H_2$) were determined for the supported iron oxide samples and for several supported zinc oxide samples. The coordinated use of infrared spectroscopy and Mössbauer spectroscopy allows the observation of both the solid state and adsorptive properties of the samples which may then be related to the water-gas shift activities of these materials.

The supports employed in this study were SiO_2, $\gamma-Al_2O_3$, TiO_2, MgO, and ZnO. These oxides were chosen as supports because they display a wide range of acid/base properties, i.e., SiO_2 is acidic, Al_2O_3 and TiO_2 are amphoprotic, and MgO and ZnO are basic. The loading of iron and zinc on the support materials was varied from 1 to 50 cation %. On the low loading samples, the supported cations are highly dispersed on the support materials, and the interaction between the cations and the support is expected to be enhanced. Higher loading samples were studied to determine if supported, bulk phase iron oxide and zinc oxide have the catalytic properties of the unsupported oxides or the properties of the highly dispersed, lower loading samples.

Mössbauer spectroscopic studies of supported iron have previously been carried out for Al_2O_3 (15-18), SiO_2 (16,17,19-22), and MgO (16,23-25). Those studies were typically conducted either under reducing conditions or in air. For the results reported in this paper, the supported iron samples were studied in CO/CO_2 gas mixtures in which the ratio of CO to CO_2 in the pretreatment gas was varied from oxidizing to reducing conditions. The nature of the interaction between the iron cations and the support was then probed by monitoring changes in the Mössbauer spectra as the pretreatment conditions were altered.

Nitric oxide was chosen as the adsorption probe for infrared spectroscopy because studies by others indicate that this molecule is a selective probe for iron cations (22,26-31). Volumetric adsorption of NO has been used to determine the dispersion of supported iron samples (27,32). In the present work, infrared bands for adsorbed NO are shown to distinguish between different types of iron surface sites.

Experimental

Sample Preparation   The supports employed in this study were $\gamma-Al_2O_3$ (Davison, SMR 7-5913), SiO_2 (Davison, Grade 952), ZnO (New Jersey Zinc, CO-064), TiO_2 (Degussa, P25), and MgO (prepared by precipitation and decomposition of magnesium hydroxide, as described elsewhere (24)). The preparation of these samples was presented in

detail elsewhere (33). Briefly, the samples were prepared by
incipient wetness impregnation of the supports with an aqueous
solution containing the proper concentration of $Fe(NO_3)_3 \cdot 9H_2O$ or
$Zn(NO_3)_2 \cdot 6H_2O$. Because Mössbauer spectroscopy for iron is sensitive
only to ^{57}Fe, and the natural abundance of ^{57}Fe is low (ca. 2%), the
samples used for Mössbauer spectroscopy studies were prepared from
isotopically enriched iron salts as described elsewhere (33). The
amount of solution used per gram of support material, the loadings
as determined by Galbraith laboratories, and the BET surface areas
and average particle diameters of the samples following water-gas
shift kinetics are listed in Table I. Particle diameters were
determined by X-ray diffraction line broadening according to the
procedure described by Klug and Alexander (34). The dispersion of
iron cations on these samples as determined by the extent of NO
adsorption is also indicated in Table I.

Mössbauer Spectroscopy Mössbauer spectroscopy studies were carried
out using the apparatus discussed by Phillips et al. (35). Doppler
velocities were calibrated using sodium nitroprusside or a
12.5 μm metallic foil at room temperature. All isomer shifts are
given relative to metallic iron. Samples were prepared for
Mössbauer spectroscopy study by compressing about 200 mg of material
at 34 MPa into a 2.54 cm diameter wafer. The effect of varying the
CO/CO_2 ratio on the oxidation state of supported iron cations was
investigated using mixtures containing CO and CO_2 in the ratios
40/60, 15/85, and 1/99. Several samples were treated in flowing
hydrogen.

Infrared Spectroscopy Infrared spectra were collected in the
transmission mode using a Nicolet 7199 FTIR instrument with an MCT
detector. Samples containing 20 mg/cm^2 were placed in a stainless
steel cell which allowed the collection of in situ spectra over the
temperature range from 300 to 700 K and in the pressure range from
vacuum to 0.2 MPa. A mixture of $CO/CO_2/He$ (2/11/87, from Matheson,
99.95% pure) was used for sample treatments without further
purification. The samples were then exposed to NO at a pressure of
20 kPa. A detailed description of the cell and the instrumental
settings used are given elsewhere (36).

Water-Gas Shift Kinetics Water-gas shift kinetics were measured
using the apparatus described by Lund and Dumesic (5). Each sample
was pretreated for 4 or more hours at 660K in a flowing mixture of
CO/CO_2 (15/85) before reaction kinetics were initiated. Reaction
kinetics were measured using a synthesis gas stream prepared by
flowing a mixture of CO/CO_2 (89/11) through a water saturator at
375 K. This yielded a "standard" synthesis gas with the following
partial pressure: 32 kPa CO, 4 kPa CO_2, and 64 kPa H_2O. After
switching to synthesis gas, catalytic activity was monitored for
24 h to verify that stable activity had been obtained before
reaction kinetics studied began.
    The concentration dependence of the reaction kinetics was
modeled using the power law expression described by Bohlbro
(37) ($-r_{CO} = k_{CO}P_{CO}^{\ell}P_{H_2O}^{m}P_{CO_2}^{q}$). Only CO, CO_2, and H_2O were included in

TABLE I

Surface area, NO uptake, and dispersion of the supported oxide catalysts

Sample	Solution Volume[a] (cm³/g)	Loading[b] (cation %)	Surface Area (m²/g)	NO Uptake (μmol/g)	Dispersion (nm)	$\bar{D}$[c]
1%Fe/Al$_2$O$_3$	0.85	0.95	224	28	0.15	—
10%Fe/Al$_2$O$_3$	0.85	—	175	312	0.17	—
25%Fe/Al$_2$O$_3$	0.85	—	138	468	0.11	12
1%Fe/TiO$_2$	0.50	1.23	39	34	0.23	—
25%Fe/TiO$_2$	0.50	—	22	150	0.05	23
1%Fe/SiO$_2$	2.0	1.27	261	40	0.24	—
25%Fe/SiO$_2$	2.0	—	194	56	0.015	20
50%Fe/SiO$_2$	2.0	—	130	143	0.019	—
1%Fe/MgO	0.75	0.76	43	48	0.27	—
1%Fe/ZnO	0.40	1.34	—	—	0.2[d]	—

a. Amount of solution used to impregnate the support
b. Based on cations (as determined by Galbraith Laboratories)
c. Average particle diameter from x-ray line broadening
d. By analogy to iron cations on other supports at same loading

the feed stream. Work by others indicates that the reaction is zero
order in H_2 (37).

Turnover frequencies were calculated for the following standard
set of conditions: T = 653 K, P_{CO} = 32 kPa, P_{CO_2} = 4 kPa, and P_{H_2O} =
64 kPa. These calculations were based on site densities calculated
from the NO uptakes, assuming one iron cation per adsorbed NO
molecule. Because zinc cations do not adsorb NO it was impossible
to directly measure the dispersion of supported zinc cations.
Therefore, to a first approximation the dispersions of the supported
zinc oxide samples were assumed to be same as the dispersions of
analogous supported iron oxide samples. The water-gas shift
activities of the iron and zinc cations supported on Al_2O_3, TiO_2,
and SiO_2 were corrected for support activity by subtracting the
activities of the supports from the total activities of the
supported catalyst. On these samples, the activities of the
supports accounted for 5-30% of the overall catalytic activity. For
Fe/MgO and Fe/ZnO the activities of the supports were, within
experimental error, equal to the activities of the supported
samples, and the activity of the supported iron cations could not be
determined.

Gases   All CO/CO_2 mixtures used in this study (premixed from
Matheson, 99.5% pure) were passed through a bed of glass beads at
620 K prior to use. This decomposed any metal carbonyl species
which may have been present in these gases. Carbon dioxide was
obtained from Liquid Carbonic (99.995% pure) and was used without
further purification. Nitric oxide was obtained from Matheson and
was purified by passing it through silica gel at 77 K prior to
use. Hydrogen (Air Products 99.9% pure) was purified by flowing
through a Deoxo unit (Englehard) followed by an activated molecular
sieve trap (13 X) at 77 K.

Results

Mössbauer Spectroscopy   The Mössbauer spectral characteristics for 1%
loadings of iron cations on the supports are summarized in Table
II. Detailed results of the Mössbauer spectroscopy studies are
presented elsewhere (33). Typical room temperature Mössbauer
spectra are shown in Figure 1. Following treatment in CO/CO_2
(15/85) at 660 K for about 16 h, spectra 1a and 1b were collected
for 1% Fe/Al_2O_3 and 1% Fe/SiO_2, respectively. The spectrum for 1%
Fe/Al_2O_3 (spectrum 1a) was fit with two overlapping doublets typical
of Fe^{2+} (see Table II). The relative areas of the doublets could be
varied while obtaining an equally good fit, suggesting that the
sample surface was heterogeneous, with a distribution of doublets
superimposed to give the observed broad doublet. Indistinguishable
spectra were obtained following treatment for 16 h at 660 K in
CO/CO_2 (1/99) or CO/CO_2 (40/60).

The Mössbauer spectrum for 1% Fe/SiO_2 (spectrum 1b) is similar
to those obtained by Yuen et al. (22) and was fit with three Fe^{2+}
doublets. The two outermost doublets (which had parameters similar

TABLE II

Mössbauer Parameters for Supported Iron Oxide Samples

Sample	Fe^{2+} (outer)			Fe^{2+} (inner)			Fe^{3+}		
	I.S. (mm/s)	Q.S. (mm/s)	Area (%·mm/s)	I.S. (mm/s)	Q.S. (mm/s)	Area (%·mm/s)	I.S. (mm/s)	Q.S. (mm/s)	Area (%·mm/s)
1% Fe/Al$_2$O$_3$	1.07	2.16	6.7	-	-	-	-	-	-
	0.99	1.50	8.1	-	-	-	-	-	-
1% Fe/TiO$_2$	1.04	2.16	7.7	-	-	-	-	-	-
	0.99	1.59	8.3	-	-	-	-	-	-
1% Fe/MgO	0.94	1.91	6.0	1.03	0.72	5.3	0.38	0.75	13.7
1% Fe/SiO$_2$	1.08	2.09	4.9	0.76	0.83	8.4	-	-	-
	0.99	1.65	7.1						

Note: IS = Isomer shift relative to Fe^0

QS = Quadrupole splitting

Area = Integrated area under each doublet

to those for 1% Fe/Al_2O_3) were assigned to sites of higher symmetry
than the inner Fe^{2+} doublet which had a smaller isomer shift and
quadrupole splitting (see Table II). The site assignments are
discussed in more detail below.

The Mössbauer spectroscopy results can be summarized in the
following manner. On each of the 1% loading samples studied, a
broad Fe^{2+} doublet with an isomer shift of ca. 1 mm/s and a
quadrupole splitting of ca. 2 mm/s was observed. On the Al_2O_3,
TiO_2, and SiO_2 supported samples only Fe^{2+} was observed after
treatment in any CO/CO_2 mixture. The oxidation state of iron on 1%
Fe/Al_2O_3 or 1% Fe/SiO_2 was not affected by treatment in pure CO_2;
however, treatment of 1% Fe/TiO_2 in CO_2 did partially reduce a

sample which had been oxidized to Fe^{3+}. The reducing agent may have
been a trace CO impurity in the CO_2 (33). On spectra for 1% Fe/SiO_2
an inner Fe^{2+} doublet with a quadrupole splitting of 0.7 mm/s and an
isomer shift of 1 mm/s was observed in addition to the doublet
discussed above. The room temperature Mössbauer spectra for 25%
Fe/Al_2O_3 and 25% Fe/SiO_2 treated at 660K in CO/CO_2(15/85) were fit
with two Fe^{2+} doublets having Mössbauer parameters similar to those
for the 1% loading samples and two sextuplets with parameters
characteristic of magnetite (33). A singlet characteristic of bulk
phase $FeTiO_3$ was observed for 25% Fe/TiO_2 (33).

For 1% Fe/MgO the ratio of surface Fe^{2+} to Fe^{3+} was dependent
on the CO/CO_2 ratio. Treatment of Fe/MgO in H_2 did not reduce all
Fe^{3+}, and no metallic iron was formed; however, the amount of Fe^{2+}
in bulk MgO sites was increased. Spectra for 1% Fe/ZnO were similar
to those for 1% Fe/MgO except for the presence of a sextuplet in the
Fe/ZnO spectrum.

X-Ray Diffraction   X-ray analysis was carried out on the 25%
Fe/Al_2O_3, 25% Fe/TiO_2, and 25% Fe/SiO_2 samples. The x-ray
diffraction patterns of the alumina and silica supported samples
were characteristic of magnetite. Line broadening calculations
showed that the magnetite particles had a diameter of 12 nm on 25%
Fe/Al_2O_3, and 20 nm on 25% Fe/SiO_2. The absence of peaks for the
support material indicated that the Al_2O_3 and SiO_2 were either
amorphous or existed as small particles (< 5 nm).

The x-ray diffraction pattern of the 25% Fe/TiO_2 sample
indicated the presence of $FeTiO_3$ and the anatase and rutile crystal
structures of TiO_2 (38). The particle diameter of the iron titanate
phase was calculated to be ca. 23 nm. The 25% Fe/TiO_2 sample had a
higher dispersion than the 25% Fe/Al_2O_3 sample (see Table I) yet the
particle size of the iron-containing phase is similar in both
cases. This suggested that on 25% Fe/TiO_2, in addition to the bulk
iron titanate phase, iron cations were also dispersed over the TiO_2
support.

Infrared Spectroscopy   The infrared spectroscopy bands observed
following adsorption of NO at beam temperature (ca. 310 K) are given
in Table III. Detailed results of the IR spectroscopy studies are
presented elsewhere (39). Typical results are shown in Figures 2
and 3. These are the beam temperature spectra of 1% Fe/Al_2O_3 and 1%
Fe/SiO_2 in NO. The samples were initially treated for 4 h in

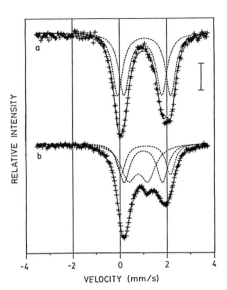

Figure 1.   Room—temperature Mössbauer spectra of 1% Fe/Al$_2$O$_3$ and 1% Fe/SiO$_2$ after treatment in CO/CO$_2$ (15/85) at 660 K for 16 h: a) 1% Fe/Al$_2$O$_3$; b) 1% Fe/SiO$_2$. The vertical bar beside the figure corresponds to a 5% change in γ—ray transmission through the sample.

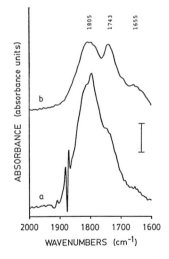

Figure 2.   Beam—temperature IR spectra of 1% Fe/Al$_2$O$_3$ after exposure to 20 kPa NO; (a) 120 min in NO at beam temperature; (b) evacuation to 10^{-2} Pa for 30 min at beam temperature.   The vertical bar beside the figure corresponds to an absorbance of 0.01.   The feature at 1870 cm^{-1} is due to incomplete subtraction of spectral features due to gaseous NO.

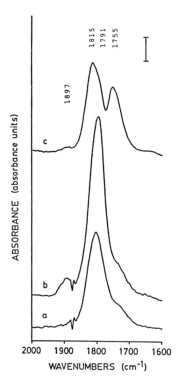

Figure 3. Beam-temperature IR spectra of 1% Fe/SiO_2 after exposure to 20 kPa NO: (a) 1 min in NO at beam temperature; (b) 130 min in NO at beam temperature; (c) evacuation to 10^{-2} Pa for 30 min at beam temperature. The vertical bar beside the figure corresponds to an absorbance of 0.01. The feature at 1870 cm^{-1} is due to incomplete subtraction of spectral features due to gaseous NO.

$CO/CO_2/He$ (2/11/87) at 660 K and evacuated to 10^{-2} Pa for 1 h at 660 K. This treatment has been shown to convert essentially all of the iron to the ferrous state. Spectrum 2a was taken following exposure of 1% Fe/Al_2O_3 to NO at beam temperature for 120 min. Gas phase NO bands and the background spectrum of the sample (which was collected before exposure to NO) were subtracted from each spectrum so that only spectral features caused by exposure to NO are displayed. Spectrum 2a was fit with infrared bands at 1805, 1743, and 1655 cm^{-1}. The intensities of these bands were essentially constant after the first two minutes of exposure to gaseous NO. The peak at 1655 cm^{-1} can be assigned to a nitrate band on Al_2O_3 (36). Evacuation to 10^{-2} Pa at beam temperature for 30 min resulted in spectrum 2b. The peaks at 1805 and 1743 cm^{-1} were reduced in intensity, although the latter peak was less sensitive to evacuation. Indeed, the peaks at 1805 and 1743 cm^{-2} are now easily discernible in spectrum 2b. Reexposure to NO and evacuation to 10^{-2} Pa reproduced spectra 2a and 2b, respectively, indicating the reversibility of this process.

TABLE III
Adsorbed NO Infrared Bands for Supported Iron Oxide Samples

Sample	Mononitrosyl (cm^{-1})			Dinitrosyl (cm^{-1})	
1% Fe/Al_2O_3		1805	1743	–	–
1% Fe/TiO_2	1838	1824	1740	–	–
1% Fe/MgO		1800	1720	–	–
1% Fe/ZnO		1813	1735	–	–
1% Fe/SiO_2		1815	1755[a]	1897	1791

a. This band is adsorbed on the same sites as the dinitrosyl species.

Beam temperature spectra of 1% Fe/SiO_2 after adsorption of NO are shown in Figure 3. After treatment in $CO/CO_2/He$ (2/11/87), as described above, the sample was exposed to NO at beam temperature for 1 min (spectrum 3a). Nitrosyl bands were observed at 1802 and 1755 cm^{-1}. Following exposure to NO for 130 min (spectrum 3b) the band at 1802 cm^{-1} had shifted to 1795 cm^{-1} and increased in intensity, the band at 1755 cm^{-1} increased in intensity, and a new band was observed at 1897 cm^{-1}. These spectra were fit using four bands. Specifically, the peak at ca. 1800 cm^{-1} was fit with two bands, one at 1815 cm^{-1} and the other at 1791 cm^{-1}, and bands were also included at 1897 and 1755 cm^{-1}.

These bands have been assigned according to Yuen et al. (22). The bands at 1815 and 1755 cm^{-1} are assigned to mononitrosyl species, while those at 1897 and 1791 cm^{-1} are assigned to the symmetric and asymmetric stretching modes, respectively, of an iron dinitrosyl species. The dinitrosyl species is converted to the mononitrosyl species at 1755 cm^{-1} by evacuation. This will be discussed in more detail below.

Evacuation at beam temperature caused a marked decrease in the bands at 1897 and 1791 cm^{-1}, and the band at 1755 cm^{-1} was reduced

slightly (see Figure 3c). Importantly, the band at 1755 cm^{-1}
increased in intensity upon evacuation. Reexposure to NO for 36 min
increased the intensities of the bands at 1897, 1815, and 1791 cm^{-1}
and decreased the intensity of the band at 1755 cm^{-1}, returning the
band intensities to their values before evacuation. A second
evacuation had the same effect as the first, indicating that this is
a reversible process.

Mononitrosyl species attributable to NO adsorbed on iron
cations were observed for all of the supported iron oxide samples
studied (see Table III). Typically one mononitrosyl species was
observed at 1850–1800 cm^{-1}, and a second mononitrosyl was observed
near 1750 cm^{-1}. These mononitrosyl species reached their maximum
intensities after exposure to NO within a few minutes. The
appearance of several mononitrosyl bands on each sample suggests
that the surfaces are heterogeneous.

The 1% Fe/SiO_2 sample displayed bands for both mononitrosyl and
dinitrosyl species. The dinitrosyl species were indicated by bands
at ca. 1900 cm^{-1} and 1800 cm^{-1} which decreased in intensity upon
evacuation, accompanied by a corresponding increase in the intensity
of the mononitrosyl peak at ca. 1750 cm^{-1}. Therefore, the
mononitrosyl band at 1750 cm^{-1} and the dinitrosyl bands are thought
to be adsorbed on the same site. A second mononitrosyl band was
observed at 1815 cm^{-1}. Unlike mononitrosyl species, for which the
maximum intensities were attained rapidly, the intensities of the
dinitrosyl species increased over a 2 h period. This is in agreement
with the work of Segawa et al. (31) for Fe^{2+} in γ-zeolite.

Water–Gas Shift Kinetics Studies   The activation energies, power law
exponents, and turnover frequencies of the supported samples for
water–gas shift are given in Table IV. The temperature range over
which the activation energies were determined and the power law
exponents are also indicated. On Fe/Al_2O_3, Fe/TiO_2 and Zn/Al_2O_3 the
water–gas shift reaction was approximately first order in CO and
about 0.25 in H_2O. These values are similar to those reported for
the bulk oxides Fe_3O_4, ZnO, and MgO (37,40). Over the bulk oxides,
CO_2 inhibited the water–gas shift reaction (40); however, on the
supported samples discussed above, the inhibition by CO_2 is absent,
as indicated by the zero order dependence on CO_2. The concentration
dependence of the rate was not determined over the other supported
samples because of low activity (Fe/SiO_2 and Zn/SiO_2) or
interference by the support (Fe/MgO and Fe/ZnO).

The activation energies of the supported oxides were lower than
those of the corresponding bulk oxides. Activation energies for
Fe_3O_4 and ZnO are typically 105–115 kJ/mol (40), while a value of
70–80 kJ/mol was observed for Fe/Al_2O_3, Zn/Al_2O_3, Fe/TiO_2, and
Zn/SiO_2. The activation energies of some materials were not
determined because of low activity (Fe/SiO_2) or support interference
(Fe/MgO and Fe/ZnO).

The following order of decreasing catalytic activity was
observed at 653 K: $Fe/Al_2O_3 \simeq Zn/Al_2O_3 \simeq Fe/TiO_2 > Fe/SiO_2 \simeq$
Zn/SiO_2. On MgO and ZnO, the activity of the support masked the
catalytic activity of the supported iron cations; therefore, it is
suggested for these samples that the activity of the supported iron
cations is comparable to, or less than that of the support.

TABLE IV

Kinetic parameters of supported iron and zinc oxides for the water-gas shift reaction including activation energy (E_A), surface area (S.A.), power law exponents (l, m, and q), and the turnover frequency at 653 K (TOF)

Sample	Activation Energy (kJ/mol)	Temp[a] Range (K)	Power Law Exponents[b]			Temp[c] (K)	Turnover Frequency (s^{-1})
			CO l	H_2O m	CO_2 q		
1%Fe/Al_2O_3	83	660-720	0.95	0.30	0.00	695	$1.*10^{-3}$
10%Fe/Al_2O_3	80	660-710	0.80	0.30	-0.05	675	$3.*10^{-4}$
25%Fe/Al_2O_3	86	630-685	0.85	0.15	-0.05	645	$3.*10^{-4}$
1%Zn/Al_2O_3	67	620-695	1.00	0.35	0.00	670	$3.*10^{-3}$
1%Fe/TiO_2	69	645-690	0.90	0.35	0.00	690	$1.*10^{-3}$
25%Fe/TiO_2	–	–	–	–	–	–	$4.*10^{-4}$
25%Fe/SiO_2	–	–	–	–	–	–	$4.*10^{-5}$
50%Fe/SiO_2	–	–	–	–	–	–	$4.*10^{-5}$
10%Zn/SiO_2	70	665-710	–	–	–	–	$8.*10^{-5}$
1%Fe/MgO	–	–	–	–	–	–	d
1%Fe/ZnO	–	–	–	–	–	–	d

a. Temperature range over which the activation energy was determined
b. Exponents l, m, and q are unitless
c. Temperature at which the exponents were determined
d. Activity of iron masked by the support

The turnover frequency for iron or zinc cations on Al_2O_3 and TiO_2 was about 10^{-3} s^{-1}. This is about two orders of magnitude lower than the activity observed for magnetite, but it is comparable to the activity of unsupported ZnO. The turnover frequency was nearly independent of iron loading, with a factor of two decrease as the loading for Fe/Al_2O_3 was increased from 1% to 10%. Both iron and zinc cations on SiO_2 had a turnover frequency of about 10^{-5} s^{-1}. Again, no dependence on iron loading was observed.

Discussion

<u>Mössbauer Spectroscopy and Infrared Spectroscopy</u> It is proposed that a significant fraction of the ferrous cations for all of the 1% loading samples is present in sites of high coordination (e.g., 5- or 6-fold). Sites of lower coordination (e.g., 3- or 4-fold) are also suggested to exist on SiO_2 and TiO_2. Finally, it is suggested that the Fe^{2+} cations on Fe/Al_2O_3, Fe/TiO_2, and Fe/SiO_2 are stabilized by the formation of a surface phase containing iron cations and cations from the support.

Mössbauer spectra of the iron containing samples indicate that the iron cations are stabilized as Fe^{2+} over Al_2O_3, TiO_2, and SiO_2 when treated in CO/CO_2 gas mixtures. The formation of a surface phase containing iron cations and cations from the support could account for the stability of Fe^{2+}. In fact, surface spinels (MAl_2O_4, where M is a divalent metal cations) have been postulated to exist for iron (41,42) and zinc (43) cations supported on Al_2O_3; and, Lund and Dumesic (5-8) have reported that a surface phase containing iron cations and silicon cations was responsible for the low water-gas shift activity of Fe_3O_4/SiO_2 samples.

Surface phases consisting of iron cations and cations from the support may also exist on samples containing 25% Fe. Mössbauer spectra of 25% Fe/Al_2O_3 indicated that about 50% of the iron was present as magnetite with the remainder present as Fe^{2+}. The turnover frequency of this sample for water-gas shift should, therefore, have been comparable to Fe_3O_4; however, the turnover frequency of this sample was two orders of magnitude lower than that of magnetite. In fact, increasing the iron loading from a 1% loading to a 25% loading on Al_2O_3, TiO_2, and SiO_2 had little affect on the turnover frequency. This suggests that the Fe_3O_4 particles on the support were covered by a surface phase. It is interesting to note that an $FeAl_2O_4$ surface phase has been proposed to stabilize the dispersion of bulk Fe_3O_4 promoted with Al_2O_3 (44).

No evidence for a surface phase which stabilized Fe^{2+} was observed on Fe/MgO or Fe/ZnO. In fact, over MgO both Fe^{2+} and Fe^{3+} were observed, and their ratio could be varied by changing the ratio of CO to CO_2 in the treatment gas. This may be related to the fact that MgO and ZnO are the least acidic of the oxides investigated as supports in this study.

The large quadrupole splittings and isomer shifts observed for the iron cations on 1% Fe/TiO_2 and 1% Fe/Al_2O_3 and the absence of dinitrosyl species suggest that in surface compounds the iron cations are present in sites of high coordination (e.g., 5- or 6-fold). In the bulk compound $FeAl_2O_4$ the iron cations are located in tetrahedral sites (45); however, surface cations are expected to be coordinatively unsaturated, and coordinatively unsaturated "tetrahedral" cations may be less stable than unsaturated "octahedral" cations. Cobalt cations supported on Al_2O_3 have, in fact, been observed to occupy tetrahedral sites in the bulk but octahedral sites on the surface (46). A doublet with similar parameters was observed in this study for all of the supported iron samples, suggesting that iron cations preferentially occupy octahedral sites or coordinatively unsaturated octahedral sites on oxide surfaces, giving rise to 6- or 5-fold coordination, respectively.

The Mössbauer spectra of Fe^{2+} on SiO_2, and the infrared spectra for NO on these samples, were similar to those reported in the literature. Two distinct types of iron sites were observed. These are characterized in the Mössbauer spectra by two doublets, a doublet with a smaller isomer shift and quadrupole splitting denoted as an "inner doublet" and a second doublet denoted as an "outer doublet". The outer doublet iron has been assigned to sites of higher coordination (e.g., 5- or 6-fold), while the inner doublet has been assigned to sites of lower coordination (e.g., 3- or 4-fold). Dinitrosyl species may be formed on the "inner doublet sites" which are lower in coordination.

In Table III it is evident that the mononitrosyl species fall into two groups, one with a stretching frequency near 1825 cm^{-1} and another at about 1750 cm^{-1}. Note that the band at 1750 cm^{-1} is in the same region as the mononitrosyl species formed upon evacuation of the dinitrosyl species associated with iron cations of low coordination. This suggests that the band at 1750 cm^{-1} corresponds to mononitrosyl species adsorbed on cations in sites of lower coordination than those resulting in the mononitrosyl species at 1825 cm^{-1}. This is in agreement with the work of other investigators, who have correlated the stretching frequency of NO with the coordination of iron cations (22,31,47). Sites of lower coordination have less steric hindrance than sites of higher coordination, thus, allowing the formation of bent nitrosyl species. In sites of higher coordination, the nitrosyl species must remain more linear due to steric hindrance by other ligands. Indeed, the wavenumber of bent species has been observed to be less than that of linear nitrosyl species (48).

The ratio of Fe^{2+} to Fe^{3+} on Fe/MgO was shown to depend on the CO/CO_2 ratio in which the sample was treated, whereas iron was stabilized as Fe^{2+} on Al_2O_3, TiO_2, and SiO_2. Thus, it may be suggested that the exposure of Fe/MgO to NO might oxidize Fe^{2+} to Fe^{3+} accompanied by the formation of N_2O, while this reaction may be negligible over iron on the other three supports. The intensity of the band at 1745 cm^{-1} decreased while that at 1800 cm^{-1} increased during prolonged exposure of Fe/MgO to NO. The site at 1745 cm^{-1} could accordingly be assigned to a mononitrosyl species adsorbed on Fe^{2+} and the band at 1800 cm^{-1} to a mononitrosyl species adsorbed on Fe^{3+}. The electron density on NO would be expected to be higher when adsorbed on Fe^{2+} than on Fe^{3+}; and, this would be reflected by a lower stretching frequency of the N-O bond since this additional electron density would be in the antibonding orbital of NO (48).

It was noted elsewhere (36) that small concentrations of dinitrosyl species may be formed on Fe/TiO_2. If this is true, then the samples can be arranged in the following order with respect to extent of formation of dinitrosyl species in the presence of NO: $Fe/SiO_2 > Fe/TiO_2 > Fe/Al_2O_3 \simeq Fe/ZnO \simeq Fe/MgO$. This order can be related to the coordination of oxygen in these supports. In the bulk structures of these supports, each lattice oxygen is coordinated to 2 cations in silica, 3 cations in titania, 3 or 4 cations in alumina, 4 cations in zinc oxide, and 6 cations in magnesia. For a cation with a given charge, the coordination number of the cation generally decreases as the coordination of oxygen decreases. This explains why ferrous cations of low coordination (giving rise to dinitrosyl bands in infrared spectra and the inner doublet in Mössbauer spectra) were present in high concentration on silica, lower concentration on titania, and absent on the other supports studied. In agreement with these ideas, Murrell and Garten (49) have observed an inner doublet in Mössbauer spectra of iron on titania at low loadings.

Water-Gas Shift Reaction   The results of the water-gas shift kinetic studies indicate that the catalytic activity of supported iron oxide is significantly lower than that of unsupported Fe_3O_4. In addition, both iron oxide and zinc oxide are two orders of magnitude less active on SiO_2 than on Al_2O_3. In general, these samples can be placed in three groups based on their water-gas shift activities: (i) those which are active (Fe/Al_2O_3, Zn/Al_2O_3, and Fe/TiO_2), (ii) those with low activity (Fe/SiO_2 or Zn/SiO_2), and (iii) those for which the support is as active or more active than the supported iron oxide (Fe/MgO and Fe/ZnO).

Two mechanisms are thought to be important for the water-gas shift reaction: a regenerative mechanism in which the catalyst surface is successively oxidized by H_2O and reduced by CO, and an associative mechanism in which adsorbed reactant species interact to form an adsorbed intermediate, generally thought to be a formate, which decomposes to water-gas shift products (50).

The regenerative mechanism for water-gas shift is thought to dominate over magnetite (40,50), while the associative mechanism is thought to be dominant over ZnO (40,51). Surface cations which can change oxidation state are required for the regenerative mechanism. The rapid electron hopping between the Fe^{2+} and Fe^{3+} cations in the octahedral sites of magnetite is thought to facilitate the regenerative mechanism. Mössbauer spectroscopy results indicate that iron cations are stabilized as Fe^{2+} in a variety of CO/CO_2 gas mixtures over Al_2O_3, TiO_2, and SiO_2. Since the cations do not readily undergo changes in oxidation state, the regenerative mechanism could not take place; therefore, for the supported iron and zinc oxides examined in this study the associative mechanism is proposed to be the dominant reaction mechanism.

Ross and Delgass (52) studied the reverse water-gas shift reaction over unsupported Eu_2O_3 and Eu_2O_3 supported on Al_2O_3 and SiO_2. As in this study, CO_2 inhibited the reaction over unsupported Eu_2O_3, while it did not inhibit the reaction over Eu_2O_3 supported on Al_2O_3 or SiO_2. Also, the activation energy decreased when Eu_2O_3 was supported. Unlike the supported iron and zinc oxide samples, no decrease in activity was observed when europium was supported on Al_2O_3 or SiO_2. Mössbauer spectroscopy of the europium indicated the presence of both Eu^{2+} and Eu^{3+}; therefore, consistent with the above discussion, a model was proposed in which the regenerative mechanism was active over both supported and unsupported Eu_2O_3. The decreased inhibition by CO_2 and the reduction in activation energy when the europium was supported were explained by the assumption that the supported Eu_2O_3 does not adsorb CO_2 as strongly.

In a study of the water-gas shift activity of bulk oxides the electronegativity scale of Zhang (53) was shown to correlate catalytic activity (40). The results of the present study for the supported samples can also be interpreted in this manner. The electronegativity of cations included in this study increase in the order $Fe^{2+} \approx Zn^{2+} < Fe^{3+} \approx Mg^{2+} < Ti^{4+} \approx Al^{3+} < Si^{4+}$ (see Table V). Supporting an oxide on a support which is more acidic (i.e., has a higher electronegativity) than that oxide should increase the acidity of the supported oxide. Both zinc and iron cations are less acidic than SiO_2, TiO_2, and Al_2O_3. Since more acidic oxides were shown to be less active for the water-gas shift reaction (40), the supported samples should decrease in activity as the support becomes more acidic.

TABLE V

Zhang electronegativity, and metal-oxygen bond strengths for the cations included in this study.

Cation	X^a	Oxides	M-O Strength[b] (kJ/mol O)
Fe^{2+}	0.39	FeO	650
Zn^{2+}	0.66	ZnO	1000
Fe^{3+}	1.31	Fe_2O_3	1200
Mg^{2+}	1.40	MgO	640
Al^{3+}	3.04	Al_2O_3	1300
Ti^{4+}	3.06	TiO_2	2000
Si^{4+}	8.10	SiO_2	3300

a. Zhang electronegativities (53)
b. Metal oxygen bond strength as calculated elsewhere (33)

Consistent with the above arguments, the activities of the supported iron oxide and zinc oxide samples decreased in the order: $Zn/Al_2O_3 \approx Fe/Al_2O_3 \approx Fe/TiO_2 > Zn/SiO_2 \approx Fe/SiO_2$. Cations on the more acidic support (SiO_2) were less active than those on amphoprotic supports (TiO_2 and Al_2O_3); and, the supported iron samples were comparable in activity to the supported zinc samples. As discussed above, low catalytic activities of iron oxide supported on SiO_2, Al_2O_3, and TiO_2 (compared to Fe_3O_4) were observed at all iron loadings investigated in this study, and this suggests that the Fe_3O_4 particles on the support were also covered by a surface phase.

As mentioned earlier, the water-gas shift activity of iron cations on ZnO and MgO was masked by the activity of the support material; therefore, the activity of the iron cations was comparable

to or less than the activity of the support cations. Since the
Zhang electronegativites of Fe^{2+} and Fe^{3+} cations are similar to
those for Zn^{2+} and Mg^{2+}, the activities of these materials for the
water-gas shift reactions should, in fact, be similar.

Iwamoto et al. (54) studied the activity of a series of metal-
ion exchanged zeolites for the water-gas shift reaction. The lower
water-gas shift activity of the acidic cations was explained in
terms of hard-soft acid/base properties. In this model, carbon
monoxide, which is a soft base, interacts more strongly with soft
acid sites. The adsorption of CO is generally considered to be the
rate controlling step in the water-gas shift reaction. Cations of
lower acidity are generally softer acids and as such adsorb CO more
readily. This would lead to higher surface concentrations of CO,
thereby increasing the water-gas shift acitivity of the sample.

In this study the surfaces of lower acidity were found to be
more active for the water-gas shift reaction, in agreement with
Iwamoto. However, a different model is proposed to explain this
result. A potential associative pathway for the water-gas shift
reaction is shown in Figure 4. In this pathway, H_2O adsorbs on an
anion vacancy and a surface oxygen (step a). Carbon monoxide
adsorbs on a coordinatively unsaturated cation to form a carbonyl
species which then reacts with a hydroxyl group to generate a
formate intermediate (steps b and c). In steps d and e, the formate
species reacts further with surface oxygen (hydroxyl groups) to form
a carbonate (bicarbonate) which decomposes to give gaseous CO_2 and a
surface oxygen (hydroxyl group). Finally, two hydrogen atoms
combine to form H_2, returning the surface to its initial state (step
f). We now suggest that the adsorption of CO is fast and reversible
and that the rate determining step is the reaction of the carbonyl
species with a surface hydroxyl to generate a formate (step b). The
rate of this reaction may be related to the metal-oxygen bond
strength on the surface as discussed by Rethwisch and Dumesic
(40). The Zhang electronegativites and the metal-oxygen bond
strengths are given in Table V. The metal-oxygen bond strength of
an acidic oxide is greater than that of a basic oxide; therefore,
hydroxides on basic oxides are expected to react more readily with
the carbonyl species, while on acidic species such as SiO_2 the
formate species may not form. In fact, in infrared studies of these
samples no formate bands were observed on Fe/SiO_2. The general
behavior of formate species on the various supported samples may be
described as follows. On a strong acid catalyst such as SiO_2 the
metal-oxygen bond is too strong and formate species cannot form
under water-gas shift conditions. On amphoprotic species (Al_2O_3 or
TiO_2) the surface oxygen is more labile and CO may react with
hydroxyl groups; however, the metal-oxygen bond is too strong for
subsequent reaction to form carbonates or bicarbonates. Over basic
oxides (MgO, ZnO, and FeO_x) the formates react further with surface
oxygen to complete the dehydrogenation of the formate by forming
carbonates.

The decrease in the inhibition of CO_2 and the decreased
activation energy can be described by a model similar to that
proposed by Ross and Delgass (52). That is, the decrease in the
inhibition of the reaction by CO_2 suggests that carbonate species
are destabilized on the supported materials. This destabilization

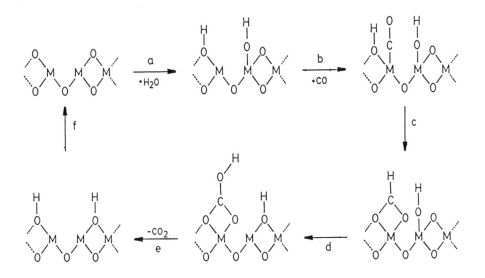

Figure 4.   Pathway for the water-gas shift reaction. (a) H_2O adsorbs on an anion vacancy and a surface oxygen; (b) CO adsorbs on anion vacancy; (c) adsorbed CO reacts with hydroxyl to form formate; (d) formate reacts to form bicarbonate; (e) CO_2 desorbs; (f) H_2 desorbs.

is accompanied by a decrease in the desorption energy of CO_2, thereby causing a reduction in the activation energy for the water-gas shift. These results can be explained by an increase in the metal-oxygen bond strength for the supported samples which makes it more difficult to form carbonate species from CO and surface oxygen. A second explanation is that because Al_2O_3 and TiO_2 are more acidic than FeO_x and ZnO, the iron and zinc cations become more acidic when supported. Since CO_2 is an acid, it should be less strongly adsorbed on an acidic oxide. Since CO_2 no longer adsorbs as readily, it no longer blocks surface sites and as a result the inhibition of the reaction by CO_2 is no longer observed.

The presence of both Fe^{2+} and Fe^{3+} on MgO suggests that the ability to alter oxidation state is not sufficient to ensure the dominance of the regenerative mechanism. The dominance of this reaction over magnetite suggests that the electron hopping which takes place in the octahedral sites facilitates the oxidation/reduction cycles necessary for the regenerative mechanism.

Conclusions

The interactions of CO and CO_2 with iron oxide and zinc oxide supported on SiO_2, Al_2O_3, TiO_2, MgO, and ZnO were examined using infrared spectroscopy, Mössbauer spectroscopy and water-gas shift kinetics. Ferrous cations supported on SiO_2, Al_2O_3, and TiO_2 were stable against oxidation to Fe^{3+} by mixtures of CO and CO_2. This stability is thought to be caused by the formation of surface compounds between iron oxide and the various supports. The low activity of supported iron oxide relative to magnetite and the stability of Fe^{2+} on the supports suggests a change in the dominant reaction mechanism for water-gas shift from the regenerative mechanism over magnetite, to the associative mechanism over the supported samples. Either CO or CO_2 may interact with surface hydroxyl species to form adsorbed formate and bicarbonate species. These surface species are proposed to be intermediates for the water-gas shift reaction. Catalytic activities of the supported samples decreased as the acidity of the support or the electronegativity of the support cations increased. Acidic supports have strong metal-oxygen bonds, and CO does not readily adsorb and react to give reaction intermediates such as formate species. Carbon dioxide adsorbs strongly on basic oxides, thereby inhibiting the water-gas shift reaction.

Acknowledgments

We are grateful to the National Science Foundation for providing the financial support for this study. Also, we would like to thank Martha Tinkle, Glenn Connell, and Luis Aparicio for valuable discussions throughout this work.

Literature Cited

1. Tauster, S.J.; Fung, S.C.; Garten, R.C. J. Am. Chem. Soc. 1978, 100, 170.
2. Tauster, S.J.; Fung, S.C.; J. Catal. 1978, 55, 29.

3. Baker, T.R.K.; Prestridge, E.B.; Garten, R.J. J. Catal. 1979, 56, 390.
4. Grasselli, R.K.; Brazdil, J.F. (eds.), Solid State Chemistry in Catalyis", ACS Symposium Series 279, American Chemical Society, 1985.
5. Lund, C.R.F.; Dumesic, J.A. J. Phys. Chem. 1982, 86, 130.
6. Lund, C.R.F.; Dumesic, J.A. J. Phys. Chem. 1981, 85, 3175.
7. Lund, C.R.F.; Dumesic, J.A. J. Catal. 1981, 72, 21.
8. Lund, C.R.F.; Dumesic, J.A. J. Catal. 1981, 76, 417.
9. Schuit, G.C.A.; Gates, B.C. AIChE J. 1973, 19, 417.
10. Massoth, F.E. Adv. Catal. 1978, 27, 265.
11. Grange, P. Catal. Rev.-Sci, Eng. 1980, 21, 135.
12. Tanabe, K.; Sumiyoshi, T.; Shibata, K.; Kiyoura, T.; Kitigawa, J. Bull. Chem. Soc. Japan 1974, 47, 1064.
13. Shibata, K.; Kiyoura, T.; Kitigawa, J.; Sumiyoshi, T.; Tanabe, K. Bull. Chem. Soc. Japan 1973, 46, 2985.
14. Tanabe, K. "Solid Acids and Bases", Academic Press, N.Y., 1970.
15. Bhide, V.G.; Date, S.K. Phys. Rev. 1968, 172, 345.
16. Hobert, H.; Arnold, D. Proc. Conf. on Appl. Moss. Effect (Tihany, 1969), 325 (1971).
17. Berry, F.J.; Maddock, A.G. Inorg. Chim. Acta. 1979, 37, 255.
18. Garten, R.L. J. Catal. 1976, 43, 18.
19. Hobson, M.C., Jr.; Gager, H.M. J. Colloid Interface Sci. 1970, 34, 347.
20. Amelse, J.A.; Butt, J.B.; Schwartz, J.H. J. Phys. Chem 1982, 86, 3022.
21. Raupp, G.B.; Delgass, W.N. J. Catal. 1979, 58, 337.
22. Yuen,S.; Chen, Y.; Kubsh, J.E.; Dumesic, J.A.; Topsøe, N.; Topsøe, H. J. Phys. Chem. 1982, 86, 3022.
23. Boudart, M.; Delbouille, A.; Dumesic, J.A.; Khammouma, S.; Topsøe, H. J. Catal. 1975, 37, 486.
24. Topsøe, H.; Dumesic, J.A.; Derouane, E.G.; Clausen, B.S.; Mørup, S.; Villadsen, J.: Topsøe, N. in "Preparation of Catalysts II, (B. Delmon, P. Grange, P.A. Jacobs, and G. Poncelet, eds.) Pg. 365, Elsevier, Amsterdam, 1979.
25. Bussiere, P.; Dutarte, R.; Martin J.P.; Mathieu, J.P. C.R. Acad. Sci. Paris Sect.6, 1975, 208C, 1133.
26. Otto, K.: Shelef, M. J. Catal. 1970, 18, 184.
27. Lund, C.R.F.; Schorfheide, J.J.; Dumesic, J.A. J. Catal. 1979, 57, 105.
28. Jermyn, J.W.; Johnson, T.J.; Vansant, E.F.; Lunsford, J.H. J. Phys. Chem. 1973, 77, 2964.
29. Rebenstorf, B. Acta. Chem. Scand. 1977, Ser. A, 31, 547.
30. Bloomquist, J.; Csillag, S.; Moberg, L.C.; Larsson, R.; Rebenstorf, B. Acta Chem. Scand. 1979, Ser. A, 33, 515.
31. Segawa, K.I.; Chen, Y.; Kubsh, J.E.; Delgass, W.N.; Dumesic, J.A.; Hall, W.K. J. Catal. 1982, 76, 112.
32. Kubsh, J.E.; Lund, C.R.F.; Chen, Y.; Dumesic, J.A. React. Kinet. Catal. Lett. 1981, 17, 115.
33. Rethwisch, D.G.; Dumesic, J.A. J. Phys. Chem. 1986, 90, 1863.
34. Klug, H.P.; Alexander, L.E. "X-ray Diffraction Procedures", 2nd ed., Wiley, New York, 1974.
35. Phillips, J.; Clausen, B.S.; Dumesic, J.A. J. Phys. Chem. 1980, 84, 1814.

36. Rethwisch, D.G., Ph.D. Thesis, University of Wisconsin, 1985.
37. Bohlbro, H., "An Investigation on the Kinetics of the Conversion of Carbon Monoxide by Water Vapour over Iron Oxide Base Catalyst", Gjellerup, Copenhagen, 1969.
38. Powder Diffraction File, Int. Centre for Diffraction Data.
39. Rethwisch, D.G.; Dumesic, J.A. J. Phys. Chem. 1986, 90, 1625.
40. Rethwisch, D.G.; Dumesic, J.A. Appl. Catal. 1986, 21, 97.
41. Hobson, M.C., Jr.; Gager, H.M. J. Catal. 1970, 16, 254.
42. La Jacono, M.; Schiavello, M.; Cimino, A. J. Phys. Chem. 1971, 75, 1044.
43. Strohmeier, B.R.; Hercules, D.M. J. Catal., 1983, 86, 266.
44. Borghard, W.S.; Boudart, M J. Catal. 1983, 80, 194.
45. Dickson, B.L.; Smith, G. The Canadian Mineralogist 1976, 14, 206.
46. Topsøe, N-Y.; Topsøe, H. J. Catal. 1982, 75, 354.
47. Tanabe, K.; Ikeda, H.; Iizuka, T.; Hattori, H. React. Kinet. Catal. Lett. 1979, 11, 149.
48. Enemark, J.H.; Feltham, R.D. Coord. Chem. Rev. 1974, 13, 339.
49. Murrell, L.L.; Garten, R.L. Appl. of Surf. Sci. 1984, 19, 218.
50. Lund, C.R.F.; Kubsh, J.E.; Dumesic, J.A. in "Solid State Chemistry in Catalysis", ACS Symposium Series 279 (R.K. Grasselli and J.F. Brazdil, eds), American Chemical Society, pg. 313 (1985).
51. Newsome, D.S. Catal. Rev.-Sci. Eng. 1980, 21(2), 275.
52. Ross, P.N., Jr.; Delgass, W.N. J. Catal. 1974, 33, 219.
53. Zhang, Y. Inorg. Chem. 1979, 21, 64.
54. Iwamoto, M.; Hasuwa, T.; Furukawa, H.; Kagawa, S. J. Catal. 1983, 79, 291.

RECEIVED May 7, 1987

Chapter 9

Adsorption and Reaction of Carbon Dioxide on Zirconium Dioxide

Ronald G. Silver, Nancy B. Jackson, and John G. Ekerdt

Department of Chemical Engineering, University of Texas, Austin, TX 78712

The activation of carbon dioxide was studied over a zirconium dioxide catalyst via infrared spectroscopy and ^{18}O-labeled reactants. The carbon dioxide adsorbed on the surface as either a carbonate or a bicarbonate species. The carbonate species formed as a result of CO_2 interaction with lattice oxygen. The bicarbonate species formed from CO_2 interaction with a hydroxyl group. There was no direct interconversion between the carbonate and the bicarbonate. It is proposed that the bicarbonate can be converted to the formate via molecular CO.

Previous reports from our laboratory (1-4) have suggested that CO is activated over zirconia via a formate and that the formate is reduced to the methoxide. Carbon monoxide hydrogenation via formate and methoxide species has also been proposed over Cu/ZnO (5). Methoxide was proposed to be the immediate precursor to methane and methanol via hydrogenation and hydrolysis, respectively (3,4). The proposed mechanism for CO interaction with bridging hydroxyl groups to form the formate and the incorporation of lattice oxygen of zirconia into the formate and methoxide species was based on oxygen labeling studies with CO and H_2O (4).

A previous study (1) has also shown that CO_2 can be converted into methane. Infrared studies have shown that CO_2 forms a bicarbonate in accordance with the adsorption studies of Tret'yakov et al. (6). Heating the bicarbonate-containing zirconia in hydrogen led to a surface methoxide species (2). It has been suggested that CO_2 interacts with terminal hydroxyl groups to form the bicarbonate (4). The route from CO_2 to methoxide (the methane precursor) has not been reported over zirconia.

Formate has been reported over ZnO from CO_2 and H_2 during the water-gas shift reaction (7,8) and following exposure of Cu/ZnO to CO_2/H_2 (9). Several groups (5,10-13) have reported the direct conversion of CO_2 into methanol. Chinchen et al. (11) proposed that CO_2 adsorbed on Cu/ZnO/Al_2O_3 and reacted with hydrogen atoms to form a

formate species, and that the water-gas shift and the CO_2-to-formate
reactions occurred by different intermediates. Vedage et al. (5)
have proposed that both CO and CO_2 react to formate and methoxide
over Cu/ZnO with the hydrogenation rate of CO much faster than CO_2.

This paper reports oxygen-labeling studies directed toward identifying the mechanisms for CO_2 interaction with zirconia. The mechanism of bicarbonate formation and its subsequent conversion to the
formate either directly or via the intermediate formation of CO were
explored.

Methods

The experimental apparatus, the catalyst synthesis from zirconium
chloride, and the catalyst characterization are reported elsewhere
(4). The experiments were conducted in a conventional fixed-bed
temperature-programmed desorption apparatus operating at 1 atm total
pressure. Water vapor was introduced to the system by sparging
helium gas through water at 25°C.

One and one-half grams of zirconia were placed in a 12.70
mm-o.d. quartz tube, and pretreated as follows: The tube and its
contents (the system) were positioned in a furnace which was heated
to 650°C in flowing oxygen and maintained at those conditions for 30
minutes. All gases were at 1 atm pressure and flowed through the
tube at a total flow of 30ml/min. While still at 650°C, the system
was flushed with helium for 15 minutes, and finally with hydrogen for
20 minutes. The system was then cooled in hydrogen to room
temperature.

For the labeling studies, the system was then ramped to 620°C at
1.0°C/sec in water-saturated helium. Next, the system was held at
620°C in water-saturated helium for 5 minutes. Dry helium cooled the
system to 450°C. A 50/50 mixture of CO_2/H_2 flowed through the system
as it cooled from 450°C to room temperature. Finally, the system was
ramped to 620°C at 1.0°C/sec in dry helium (the TPD step). Reactor
product gases were monitored for $C^{16}O$, $C^{18}O$, $C^{16}O_2$, $C^{16}O^{18}O$, $C^{18}O_2$,
$H_2^{16}O$, and $H_2^{18}O$, AMUs (atomic mass unit) 28, 30, 44, 46, 48, 18, and
20, respectively.

Additional experiments were performed using the same procedure
as above, except the system was cooled from 450°C to room temperature
in pure CO_2. Other runs were made in which the temperature at which
the system was exposed to CO_2/H_2 was varied from 400 to 600°C. Cooling in CO_2/H_2 from 450°C to 25°C enhanced bicarbonate formation and
eliminated formate/methoxide formation.

The hydrogen (Big Three UHP, 99.999%) was passed through a deoxo
cylinder and a bed of 4-Å molecular sieves to remove oxygen and
water. Carbon monoxide (Big Three UHP, 99.8%) was heated to 180°C
over molecular sieves to decompose metal carbonyls. Carbon dioxide
(Big Three UHP, 99.7%) and oxygen (Big Three UHP, 99.9+%) were passed
through beds of 4-Å molecular sieves to remove water. Helium had a
minimum purity of 99.995% and was passed through a bed of 4-Å molecular sieves to remove water. Oxygen-18 labeled CO_2 (98%) and oxygen-
18 labeled water (98%) were purchased from Cambridge Isotope Laboratories and used without further treatment.

Results

Two types of hydroxyl groups have been observed over zirconia,
bridged and terminal. Water adsorbs dissociatively over Zr cations
(14) to produce both terminal and bridged hydroxyl groups (6). The
terminal groups are less stable than the bridged groups (15) and have
been suggested as the type that exchanges with gas phase water and
that interacts with CO_2 to form the bicarbonate species (4). The
system was initially ramped and held in $H_2^{18}O$ in order to exchange
the $[^{16}O]$ terminal hydroxyl groups of zirconia with $[^{18}O]$. The exchange was incomplete during the present studies and was determined
by ratioing the total amounts of $H_2^{18}O$ and $H_2^{16}O$ that evolved during
the TPD step.

Carbon dioxide and CO evolved during the TPD labeling studies.
Species desorbing from the surface were identified on the basis of
the temperature region in which they were observed. Some of the
experiments of He and Ekerdt (1) were repeated over the catalyst used
in this study. Similar TPD results were used to assign an identity
to the desorbing/reacting species, in accordance with the assignments
of He and Ekerdt. The region between 200-260°C is proposed to be
where a carbonate desorbs. The region between 450-510°C is proposed
to be where a bicarbonate species desorbs. Finally, the region
between 580-620°C is proposed to be where formate and methoxide
species desorb.

The TPD results observed following cooling in $H_2/C^{16}O_2$ are shown
in Figure 1. Peaks for AMUs 28 and 44 were formed in the carbonate
region (225°C), and peaks for AMUs 28, 30, 44, and 46 were observed
in the bicarbonate region (480°C). Masses 18 and 20 (not shown) were
present at constant levels above the background for temperatures
greater than 450°C.

The TPD results observed following cooling in $H_2/C^{18}O_2$ are shown
in Figure 2. All the carbon oxide masses except 28 were formed
during the desorption/decomposition of the carbonate (225°C). The
carbon monoxide masses (AMUs 28 and 30) displayed peaks in the bicarbonate region (480°C), while the carbon dioxide masses (AMUs 44, 46,
and 48) appeared in the bicarbonate region as shoulders on the trail-
ing edge of carbonate peaks.

The area under each of the peaks in Figures 1 and 2, as well as
an additional experiment with $C^{16}O_2/H_2^{18}O$, are presented in Table I.
The mass fragmentation for $C^{16}O_2$ generates an AMU 28 signal that is
7% of the AMU 44 signal. The overwhelming majority of the carbonate
desorbed as carbon dioxide. The bicarbonate generated both CO and
CO_2, but more CO formed than CO_2.

The adsorption of CO and CO_2 on zirconia was also studied using
infrared spectroscopy, which provides direct evidence for surface
intermediates. The results are presented in Figure 3. Carbon monoxide formed the formate (bands at 2880, 1580, 1387, and 1360 cm^{-1})
after adsorption at 225 and 500°C, and possibly a small amount of bicarbonate (band at 1610 cm^{-1}) after adsorption at 225°C. Carbon dioxide formed the bicarbonate (bands at 1610, 1430, and 1220 cm^{-1}) and
a carbonate (band at 1335 cm^{-1}) after adsorption at 225 and 500°C.

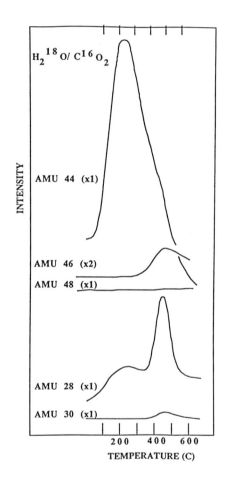

Figure 1. Mass signals during a TPD experiment following pretreatment of ZrO_2 with $H_2^{18}O$ and adsorption of $H_2/C^{16}O_2$.

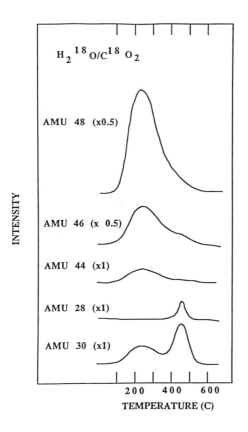

Figure 2. Mass Signals during a TPD experiment following pretreatment of ZrO_2 with $H_2^{18}O$ and adsorption of $H_2/C^{18}O_2$.

Table I. Isotope Distributions during Temperature-Programmed Heating

Preadsorbed gas (surface hydroxyl distribution)	Carbonate Peak Area AMU					Bicarbonate Peak Area AMU				
	28	30	44	46	48	28	30	44	46	48
$C^{18}O_2$/$H_2$$^{18}O$ (a) (55 : 45 ^{16}OH : ^{18}OH)	0	156	130	872	2042	45	198	7	32	55
$C^{16}O_2$/$H_2$$^{18}O$ (b) (85 : 15 ^{16}OH : ^{18}OH)	337	0	3337	0	0	584	24	187	86	6
$C^{16}O_2$/$H_2$$^{18}O$ (c) (57 : 43 ^{16}OH : ^{18}OH)	396	0	2602	17	0	395	43	203	168	15

(a) TPD shown in Figure 2
(b) TPD shown in Figure 1
(c) TPD not shown

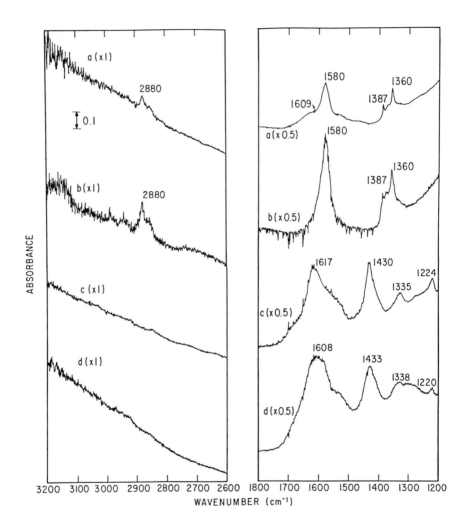

Figure 3. Infrared spectra of ZrO2 after adsorption of
(a) CO at 225°C, (b) CO at 500°C, (c) CO2 at 225°C, and
(d) CO2 at 500°C.

Discussion

Cooling the zirconia from 450 to 25°C in H_2/CO_2 resulted in the formation of carbonate and bicarbonate species. (Additional TPD studies, not shown, established that formate and methoxide species did not form during the adsorption of CO_2, under the conditions reported here.) The use of $H_2^{18}O$ to exchange ^{16}OH groups on the zirconia with ^{18}OH groups and the use of ^{18}O-labeled CO_2 allow the determination of the pathways to the formation of carbonate and bicarbonate, as well as the possible interconversion between carbonate, bicarbonate and formate. Figure 4 presents a proposed scheme for the activation of CO_2 over ZrO_2. The scheme is discussed below.

The carbonate decomposed to produce CO_2. Only $C^{16}O_2$ desorbed following the adsorption of $C^{16}O_2$ (Table I). This demonstrates that the formation of carbonate did not involve the interaction of CO_2 with the water-based hydroxyl groups of zirconia. (The water-based hydroxyl groups are most likely terminal hydroxyls (4).) A mixture of carbon dioxide isotopes was generated following the adsorption of $C^{18}O_2$. This observation suggests that carbonate is formed by the interaction of CO_2 with lattice oxygen anions of zirconia.

The infrared results with CO_2 (Figure 3) suggest that bidentate carbonate I was present. The band at 1335 cm^{-1} is typically associated with the asymmetric OCO stretch of I (16,17). Infrared studies over ZrO_2 (6) and ThO_2 (18,19) reveal that three different carbonates can form over these oxides (see Figure 4), two bidentate structures, I and II, and a monodentate structure, III. The distribution of oxygen isotopes in the TPD spectrum of CO_2 may suggest that III also formed over zirconia. If one assumes that the majority of the carbonate species formed from $C^{18}O_2$ are $[C^{16}O^{18}O^{18}O]^{-2}$, then the monodentate will produce either all $C^{18}O_2$ when the lattice oxygen $[^{16}O]$ is involved in the Zr-O-C bond, or all $C^{16}O^{18}O$ when one of the carbon dioxide's oxygens $[^{18}O]$ is involved in the Zr-O-C bond. If one assumes the same isotope composition for the bidentate carbonate, $[C^{16}O^{18}O^{18}O]^{-2}$, and that one of the Zr-O-C bonds of the bidentate carbonate is $[^{16}O]$, then the carbon monoxide that forms during carbonate decomposition should be a 1:1 mixture of $C^{16}O^{18}O$ and $C^{18}O_2$. Twice as much $C^{18}O_2$ formed as did $C^{16}O^{18}O$ during the decomposition of the carbonate, suggesting that the majority of the carbonate was present in the monodentate structure, III, with the lattice oxygen $[^{16}O]$ involved in the Zr-O-C bond. The formation of $C^{18}O^{18}O$ during carbonate decomposition is consistent with bidentate I.

The bicarbonate decomposed to produce CO and CO_2. The relative incorporation of $[^{18}O]$ into the CO and CO_2 produced following adsorption of $C^{16}O_2$ increased with the increasing $[^{18}O]$ content of the surface OH groups (Table I). This result supports the hypothesis that bicarbonate forms by the reaction between CO_2 and a terminal (water-based) hydroxyl group (4).

The labeling studies reported here as well as previous studies in our laboratory (1-4) are consistent with indirect conversion of CO_2 into C_1 hydrocarbons, via the intermediate formation of CO as shown in Figure 4. Carbon dioxide readily forms bicarbonate. The

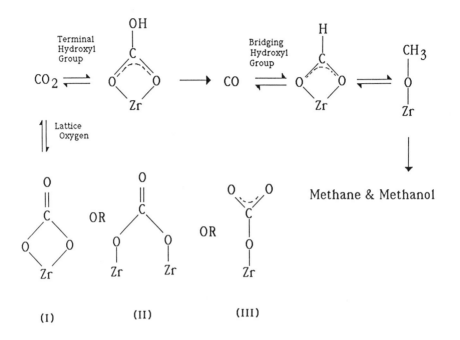

Figure 4. Proposed scheme for the activation of CO_2 over ZrO_2.

route from bicarbonate to formate remains untested. It was not possible to realize the levels of [^{18}O] incorporation into the bicarbonate and the surface concentrations of bicarbonate necessary to test for the formation of labeled-methanol from CO_2. The TPD studies have shown that bicarbonate decomposes to CO and CO_2. The infrared and other studies (4) demonstrate that CO readily reacts with ZrO_2 to produce the formate. While we cannot rule out direct reduction of the bicarbonate to a formate, the demonstrated presence of the indirect steps makes the intermediate formation of CO seem more reasonable.

Figure 4 also shows the formation of carbonates from CO_2 and a lattice oxygen. The nature of the surface sites involved in carbonate formation were not revealed in these studies. The results for [^{18}O] incorporation into the carbonate and bicarbonate species, following $C^{16}O_2$ adsorption, suggest that these species are formed at different sites and that there is no direct conversion of the carbonate into the bicarbonate.

Acknowledgments

This work was supported by the Division of Chemical Sciences, Office of Basic Energy Sciences, U.S. Department of Energy under Contract #DE-AS05-80ER10720.

Literature Cited

1. He, M-Y., and Ekerdt, J.G., J. Catal. 1984, 87, 238.
2. He, M-Y., and Ekerdt, J.G., J. Catal. 1984, 87, 381.
3. He, M-Y., and Ekerdt, J.G., J. Catal. 1984, 90, 17.
4. Jackson, N.B., and Ekerdt, J.G., J. Catal., 1986, 101, 90.
5. Vedage, G.A., Pitchai, R., Herman, R.G., and Klier, K., Proc. Int. Congr. Catal., 8th, 1984 1985, 2, 47.
6. Tret'yakov, N.E., Pozdnyakov, D.V., Oranskaya, O.M., and Filimonov, V.N., Russ. J. Phys. Chem. 1970, 44, 596.
7. Ueno, A., Yamamoto, T., Onishi, T., and Tamaru, K., Bull. Chem. Soc. Japan 1969, 42, 3040.
8. Ueno, A., Onishi, T., and Tamaru, K., Trans. Faraday Soc. 1970, 66, 756.
9. Edwards, J.F., and Schrader, G.L., J. Phys. Chem. 1984, 88, 5620.
10. Kagan, Y.B., Lin, G.I., Rozovskii, A.Y., Loktev, S.M., and Golovkin, Y.I., Kinetika i Kataliz 1976, 17, 440.
11. Chinchen, G.C., Denny, P.J., Parker, D.G., Short, G.D., Spencer, M.S., Waugh, K.C., and Whan, D.A., Preprints, ACS Division of Fuel Chemistry, Philadelphia, PA, 1984, Vol. 29, No. 5, p. 178.
12. Kung, H.H., Liu, G., and Wilcox, D., Preprints, ACS Division of Fuel Chemistry, Philadelphia, PA, 1984, Vol. 29, No. 5, p. 194.
13. Thiorolle-Cazat, J., Bardet, R., and Trambouze, Y., Preprints, ACS Division of Fuel Chemistry, Philadelhpia, PA, 1984, Vol. 29, No. 5, p. 189.
14. Agron, P.A., Fuller, E.L., Jr., and Holmes, H.F., J. Colloid and Interface Sci. 1975, 52, 553.
15. Yamaguchi, T., Nakano, Y., and Tanabe, K., Bull Chem. Soc. Japan, 1978, 51, 2482.
16. Nakamoto, K., "Infrared Spectra of Inorganic and Coordination Compounds;" Wiley: New York, 1978; p. 231.
17. Morterra, C., Coluccia. S., Ghiotti, G., and Zecchina, A., Z. Phys. Chem. 1977, 104, 275.
18. Courdurier, G., Claudel, B., and Faure, L., J. Catal. 1981, 71, 213.
19. Pichat, P., Veron, J., Claudel, B., and Mathieu, M.V., J. Chem. Phys. 1966, 33, 1026.

RECEIVED December 1, 1986

Chapter 10

Effect of Potassium on the Hydrogenation of Carbon Monoxide and Carbon Dioxide Over Supported Rh Catalysts

S. D. Worley and C. H. Dai

Department of Chemistry, Auburn University, Auburn, AL 36849

The reactions of hydrogen with carbon monoxide and carbon dioxide over Rh/Al_2O_3 and Rh/TiO_2 films, some of which contained potassium as an additive, have been investigated. For the CO hydrogenation reaction the presence of potassium caused the dissociation or desorption of the gem dicarbonyl, linear CO, and carbonyl hydride species, while it led to an enhancement of the bridged carbonyl species. For Rh/TiO_2 films the hydrogenation of CO produced acetone and acetaldehyde as oxygenated products; the bridged carbonyl species was the likely precursor of these products. For the CO_2 hydrogenation reaction the presence of potassium caused the dissociation or desorption of all CO species, and oxygenated products were not produced. Potassium significantly poisoned both reactions toward the production of methane.

Considerable effort has been expended here toward the investigation of the CO and CO_2 hydrogenation reactions over supported Rh catalysts (1-4). Although for Rh/X (X = Al_2O_3 and TiO_2) at low pressure (50-100 Torr) and temperature (423-473 K) the predominant product for both of these reactions is methane (1-4), there are means of altering the product distribution toward higher hydrocarbons and/or oxygenated products such as methanol. In general it is believed that methane and higher hydrocarbons are produced from dissociated CO or CO_2 on supported transition metals, while undissociated CO is thought to be a precursor to oxygenated products (5). Basic support materials such as ZnO, MgO, La_2O_3, and ZrO_2 (6,7) as well as higher pressure (over 1 atm) (8,9) seem to shift the product selectivity toward oxygenates; however, Al_2O_3 and TiO_2 as support materials can also lead to oxygenated products under special conditions. For example, Goodwin and coworkers (5) have shown recently that added potassium causes the selectivity for hydrogenation of CO over 3% Rh/TiO_2 to shift toward oxygenated products with acetaldehyde and acetone being

0097–6156/88/0363–0133$06.00/0

present in significant quantities. There has been considerable recent interest in the effects of alkali metals on catalytic reactions over single crystals (10) and over supported transition metals (5,11). This paper will report the results of recent work in these laboratories concerning the effects of potassium on the CO and CO_2 hydrogenation reactions over supported Rh catalytic films.

Experimental

The Rh/Al_2O_3 and Rh/TiO_2 catalysts used in this study were prepared in a manner similar to those studied previously here (1-4,12). Briefly, acetone/water solutions containing appropriate amounts of $RhCl_3 \cdot 3H_2O$, KCl, and alumina (Degussa Aluminum Oxide C, 100 m^2g^{-1}) or titania (Degussa Titanium Dioxide P25, 50 m^2g^{-1}) were sprayed using a specially designed atomizer onto a heated 20 mm CaF_2 infrared window. Evaporation of the solvents left a uniform thin film (typically 4.3 mg cm^{-2}) of the mixed solid materials adhered to the window. The window containing the film was mounted inside an infrared cell reactor (2-4) which was evacuated overnight. The sample film was then evacuated at 470 K for 1 hr, reduced at 480 K by 85 Torr doses of hydrogen for 5, 5, 10, and 20 min periods (each period followed by evacuation to ca. 10^{-5} Torr), and then evacuated for an additional hour at 480 K to a base pressure of 10^{-6} Torr. For a typical experiment the cell was then exposed to a $CO:H_2$ or $CO_2:H_2$ mixture (1:4) at ca. 82.5 Torr total pressure and heated rapidly to some prescribed temperature. Methane gas and surface intermediate formations during the reactions were monitored by infrared spectroscopy (Perkin Elmer 983 with data system) (1-4); product distributions at the end of the experiment were measured by gas chromatograpy (Carle 400). Pressure measurements were made with an MKS Baratron capacitance manometer (±0.01 Torr).

Results and Discussion

CO Hydrogenation

The interaction of CO with supported Rh catalysts has been well characterized by infrared spectroscopy (13). The primary surface species obtained are shown below. The "gem dicarbonyl" species (I) exhibits two sharp infrared bands at ca. 2030 and 2100 cm^{-1}

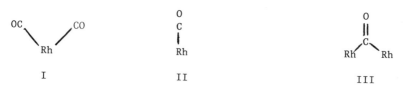

which do not generally shift in wavenumber with changing surface coverage. This indicates that dipolar coupling of nearby adsorbed CO molecules is minimal for supported Rh containing predominantly this species. For Rh/X (X = Al_2O_3, TiO_2, or SiO_2) catalysts containing less than 1% by weight Rh, species I is the only species detected by infrared. The facts that the infrared bands for species I do not shift with coverage and that no other species are detected for catalysts containing low Rh loading have led some workers to suggest that species I refers to Rh in a highly dispersed state, possibly even isolated Rh atoms (12,13). Work in several laboratories (12-14) has established that Rh in species I exists in the +1 oxidation state; this probably also contributes to its tendency to remain highly dispersed. Recent work has shown that the dispersion of Rh ions in species I may be actually caused by the presence of adsorbed CO (15,16).

Figures 1 and 2 illustrate the effect of potassium on species I for a 0.5% Rh/Al_2O_3 catalyst film. In this experiment the two catalysts were treated identically. They were held successively at temperatures of 300, 320, 380, 430 and 460 K for 30 min at each temperature in the presence of 1 x 10^{-3} Torr of CO. The infrared band intensities were very similar through 380 K, probably indicating comparable CO coverages for the two samples. However, after 30 min at 430 K species I disappeared for the sample containing potassium (K:Rh = 2:1), but this phenomenon did not occur until after 30 min heating at 460 K for the sample which did not contain potassium. Clearly potassium did not significantly block species I sites; rather it functioned to aid either CO bond dissociation or CO desorption from the surface, most probably through an electronic effect. The potassium may well be located on the support in close proximity to Rh^+ sites. Some workers have observed K/CO interactions on single crystals (17). Such interactions can give rise to low frequency CO infrared bands (1400-1800 cm^{-1}). A comparison of Figs. 1 and 2 indicates that such interactions do not occur appreciably for 0.5% Rh/Al_2O_3 catalysts which suggests that the K/CO interaction which causes enhanced dissociation or desorption of CO on or from Rh in species I may be one of long range.

When Rh/X catalysts containing higher Rh loadings (>1%) are employed, in addition to species I generally two other CO species can be detected by infrared. Species II, the "linear" species, contains one CO molecule adsorbed on a Rh atom, while species III, the "bridged carbonyl" species, contains CO bridged across two Rh atoms (13). The infrared bands for these two species shift to higher wavenumber as the CO surface coverage is increased leading most workers to suggest that these two species correspond to clusters of Rh atoms rather than highly dispersed ions as for species I. Figure 3 shows a comparison of the behavior of the CO hydrogenation reaction over 2.2% Rh/TiO_2 catalyst films with and without the presence of potassium. In spectrum 3a for a sample containing no potassium the four infrared bands normally observed are in evidence; the 2100 and 2030 cm^{-1} bands correspond to species I, while the 2072 cm^{-1} band and the broad band near 1900 cm^{-1} can be assigned to species II and III, respectively. Upon heating at 440 K for 5.5 hr (spectrum 3b), species I and II disappeared, a new band was detected at 2047 cm^{-1}, the species III band declined in intensity, and bands corresponding to gas phase methane (3015, 1304 cm^{-1}) and carbon dioxide (2349 cm^{-1}) were produced. The band at 2047 cm^{-1} corresponds to a carbonyl hydride species rather than species II. The carbonyl hydride

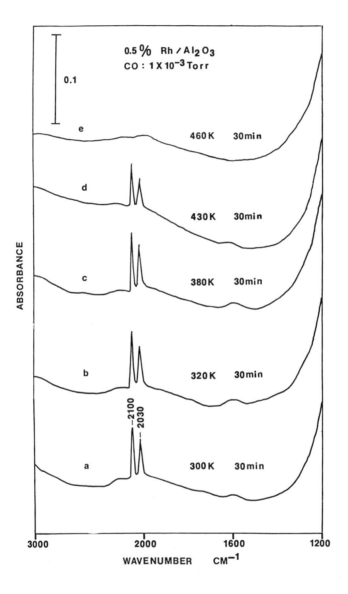

Figure 1. Infrared spectra for the interaction of CO with a 0.5% Rh/Al_2O_3 film (4.0 mg cm^{-2}) at a background pressure of 1 x 10^{-3} Torr as a function of temperature.

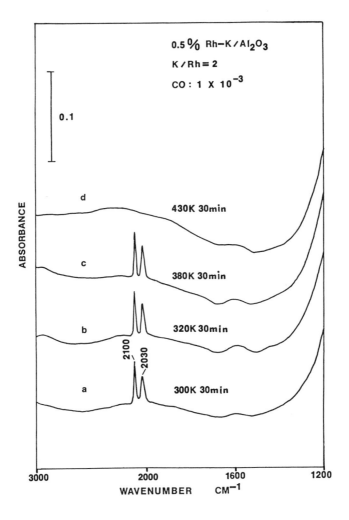

Figure 2.   Infrared spectra for the interaction of CO with a 0.5% Rh/Al$_2$O$_3$ film (4.0 mg cm^{-2}) to which has been added potassium at a K:Rh ratio of 2:1 at a background pressure of 1 x 10^{-3} Torr as a function of temperature.

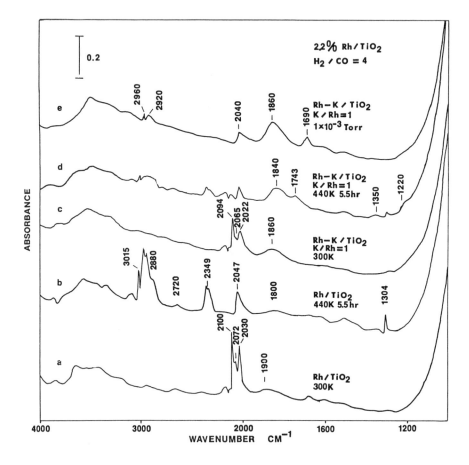

Figure 3. Infrared spectra for the interaction of H_2 and CO over 2.2% Rh/TiO_2 films (4.3 mg cm^{-2}) with or without potassium added as indicated; total pressure was 82.5 Torr.

species for supported Rh was first suggested by Solymosi and coworkers (18) and later confirmed by isotopic labeling studies here (3,4). Upon addition of potassium at 300 K (Fig. 3c), all infrared bands except the broad band for species III declined in intensity. Upon heating at 440 K for 5.5 hr (Fig. 3d), the infrared bands for species I, II, the carbonyl hydride, methane, and carbon dioxide all declined markedly in intensity relative to the identical treatment for the sample containing no potassium. Only the species III band became considerably more intense. Weak new bands at 1743, 1350, and 1220 cm^{-1} which vanish upon evacuation (Fig. 3e) can be atributed to the formation of acetone; this has been confirmed by gas chromatography. Table I shows the product distributions for these experiments. Acetone and acetaldehyde were produced as oxygenated products over Rh/TiO$_2$ and to a greater extent when potassium was present. Similar results were observed for 2.2% Rh/Al$_2$O$_3$, ie. enhanced species III (although to less extent than noted for Rh/TiO$_2$), less species I, II, and carbonyl hydride when potassium was present; however, oxygenated products were not detected for Al$_2$O$_3$ as the support. Goodwin and coworkers (5) have also observed the production of comparable amounts of acetone and acetaldehyde over potassium-doped Rh/TiO$_2$ (K:Rh = 1:2) although their reaction conditions were quite different (523 - 708 K, 1 - 10 atm, CO:H$_2$ = 2:1) than those here.

It is generally thought that oxygenated products in the CO hydrogenation reaction result from reaction of undissociated CO with hydrocarbon fragments (5). Since the presence of potassium appears to enhance the formation of the bridged carbonyl species III at the expense of the other species, it is likely that the bridged species is the precursor to the oxygenated products. The gem dicarbonyl species I and linear CO species II most likely dissociate at low temperature to form carbon and the carbonyl hydride species when hydrogen is present. The hydride ligand causes further dissociation of CO due to back donation of electron density into the 𝜋* orbital of CO. Hydrogenation of active carbon then leads to the production of methane and higher hydrocarbons. Potassium must function in several roles. It poisons the methanation reaction either by site blockage for species I, II, and the carbonyl hydride or by accelerating the production of inactive carbon through an electronic effect (electron donation into the 𝜋* orbital of CO). Furthermore, it accentuates the production of species III probably by steric blockage of species I and II thus enhancing the concentration of undissociated CO which can form oxygenated products.

CO$_2$ Hydrogenation

The hydrogenation of CO$_2$ does not produce CO species I, II, and III in infrared detectable amounts. It does produce the carbonyl

Table I. Product Distribution for CO Hydrogenation.

	$TN(CH_4)^a$	Mole Percents					
		CO_2	CH_4	C_2H_6	C_nH_m n>2	CH_3CHO	CH_3COCH_3
2.2%Rh/Al$_2$O$_3$	1.23×10^{-4}	15.8	64.0	10.4	9.9	----	----
2.2%Rh-K/Al$_2$O$_3$ K/Rh=1	1.63×10^{-5}	33.3	66.6	----	----	----	----
2.2%Rh/TiO$_2$	4.20×10^{-4}	32.0	51.3	4.7	2.4	6.7	2.9
2.2%Rh-K/TiO$_2$ K/Rh=1	4.92×10^{-5}	25.7	42.8	8.6	0.1	10.8	12.2

a Turnover number for production of methane (molecules/Rh atom/sec).

hydride species thus proving that CO_2 is dissociated over supported Rh (2-4). Figure 4 illustrates the behavior of a 2.2% Rh/Al_2O_3 catalyst for a series of experiments in which samples having different potassium loadings are heated at 440 K for 5.5 hr. Note that the infrared band at 2020-2040 cm^{-1} for the carbonyl hydride species is absent for all spectra shown; it was present at 300 K for the sample having no potassium, but it was greatly diminished in intensity at 300 K for the samples containing potassium, and it vanished at 300 K when the K:Rh ratio was 1.0. It is evident in the spectra shown in Fig. 4 that the amount of methane produced declined as the K:Rh ratio was increased; also, there was no spectroscopic evidence for oxygenated products.

Figure 5 shows the analogous results for CO_2 hydrogenation over a 2.2% Rh/TiO_2 film which contained either no potassium (Fig. 5a,b) or K:Rh = 2.0 (Fig. 5c,d). The carbonyl hydride species was present (2037 cm^{-1}) for the sample containing no potassium, but it was absent even at room temperature for the sample containing potassium. The methanation activity was substantially decreased when potassium was present, and no oxygenated products were detected by infrared or by gas chromatography (Table II). The data in Table II show that the turnover number for methane production for CO_2 hydrogenation was higher in all cases for Rh/TiO_2 than for Rh/Al_2O_3, and that potassium poisoned the methanation reaction for both types of supported catalysts. Figure 6 shows that the poisoning effect was linear for both supports, but a larger slope illustrates that the effect of potassium as an additive was more dramatic for TiO_2. The poisoning effect was exponential as a function of potassium loading for 0.5% Rh/Al_2O_3.

Infrared spectroscopy indicates that there is very little CO adsorbed intact on Rh during the hydrogenation of CO_2. Evidently the CO_2 dissociates rapidly on the supported Rh catalysts to produce active carbon which is subsequently hydrogenated to methane and a smaller amount of higher hydrocarbons. The presence of potassium during the hydrogenation of CO_2 does not cause an enhanced formation of bridged carbonyl species III as was the case for CO hydrogenation. Thus no oxygenated products are observed. Potassium probably poisons the methanation reaction by site blockage and by enhancing dissociation of CO_2 to inactive carbon.

Conclusions

Potassium poisons the methanation reaction for both CO and CO_2 hydrogenation over supported rhodium. For CO hydrogenation it causes an enhanced amount of bridged species III which is not dissociated and is probably the precursor of the oxygenated products acetone and acetaldehyde. For CO_2 hydrogenation CO species can not be detected during the reaction when potassium is present, and oxygenated products are not detected. The effect of potassium is more dramatic for CO_2 hydrogenation over Rh/TiO_2 than over Rh/Al_2O_3.

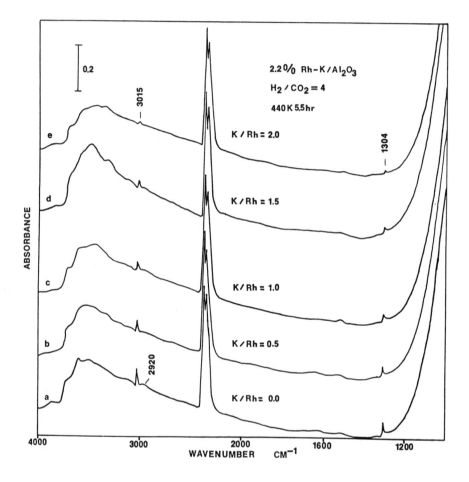

Figure 4. Infrared spectra for the interaction of H_2 and CO_2 over 2.2% Rh/Al_2O_3 films (4.3 mg cm^{-2}) as a function of concentration of potassium as an additive; total pressure was 82.5 Torr.

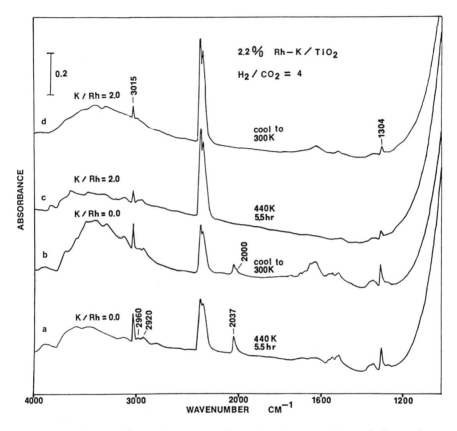

Figure 5. Infrared spectra for the interaction of H_2 and CO_2 over 2.2% Rh/TiO_2 films (4.3 mg cm^{-2}) with or without potassium added as indicated; total pressure was 82.5 Torr.

Table II. Product Distribution for CO_2 Hydrogenation.

	2.2% Rh-K/Al$_2$O$_3$			2.2% Rh-K/TiO$_2$		
	CH_4	C_2H_6	TN(CH_4)[a]	CH_4	C_2H_6	TN(CH_4)[a]
K/Rh=0.0	89.0[b]	11.0[b]	3.61×10^{-4}	97.6[b]	2.4[b]	1.48×10^{-3}
K/Rh=0.5	87.3	12.4	2.09×10^{-4}	96.7[b]	3.3	1.37×10^{-3}
K/Rh=1.0	86.4	13.6	1.62×10^{-4}	95.9	4.0	9.50×10^{-4}
K/Rh=1.2	ND	ND	ND	94.5	5.5	7.98×10^{-4}
K/Rh=1.5	86.4	13.6	1.49×10^{-4}	92.0	8.0	4.32×10^{-4}
K/Rh=2.0	ND	ND	5.82×10^{-5}	ND	ND	2.16×10^{-4}

[a]Turnover number for production of methane (molecules/Rh atom/sec).

[b]Mole percents.

ND No determination.

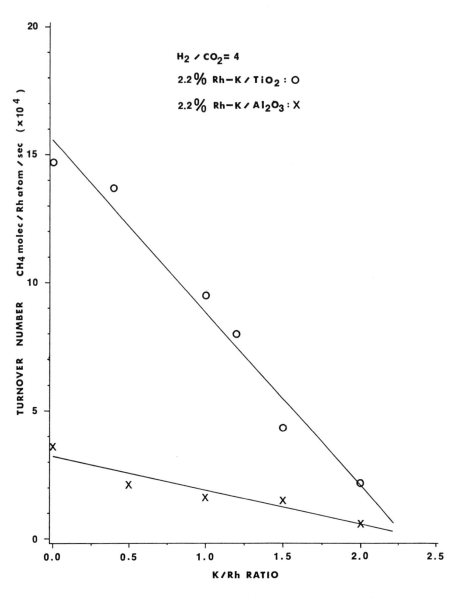

Figure 6.   Turnover number for CH_4 production as a function of concentration of potassium as an additive over 2.2% Rh/X (X = Al_2O_3, TiO_2) films (4.3 mg cm^{-2}); total pressure was 82.5 Torr, and reaction temperature was 440 K.

Acknowledgments

The authors gratefully acknowledge the support of the Office of Naval Research for this work.

Literature Cited

1. Worley, S. D.; Mattson, G. A.; Caudill, R. J. Phys. Chem. 1983, 87, 1671.
2. Henderson, M. A.; Worley, S. D. Surface Sci. 1985, 149, L1.
3. Henderson, M. A.; Worley, S. D. J. Phys. Chem. 1985, 89, 393.
4. Henderson, M. A.; Worley, S. D. J. Phys. Chem. 1985, 89, 1417.
5. Chuang, S. C.; Goodwin, J. G.; Wender, I. J. Catal. 1985, 95, 435.
6. Kawai, M.; Uda, M.; Ichikawa, M. J. Phys. Chem. 1985, 89, 1654.
7. Ramaroson, E.; Kieffer, R.; Kiennemann, A. J. Chem. Soc. Chem. Commun. 1982, 645.
8. Arakawa, H.; Fukushima, T.; Ichikawa, M.; Takeuchi, K.; Matsuzaki, T.; Sugi, Y. Chem. Lett. 1985, 23.
9. Poutsma, M. L.; Elek, L. F.; Ibarbia, P. A.; Risch, A. P.; Rabo, J. A. J. Catal. 1978, 52, 157.
10. For a few examples, see Somorjai, G. A. Catal. Rev. Sci. Eng. 1978, 18, 173; Somorjai, G. A. ibid 1981, 23, 189; Mross, W. D. ibid 1983, 25, 591; Peebles, D. E.; Goodman, D. W.; White, J. M. J. Phys. Chem. 1983, 87, 4378; Crowell, J. E.; Tysoe, W. T.; Somorjai, G. A. ibid 1985, 89, 1598.
11. For a few examples see Solymosi, F.; Tombacz, I.; Koszta, J. Catal. 1985, 95, 578, Gajardo, P.; Michel, J. B.; Gleason, E. F.; McMillan, S. J. Chem. Soc. Faraday. Diss. 1979, 72, 72; Kagami, S.; Naito, S.; Kikuzono, Y.; Tamaru, K. J. Chem. Soc. Chem. Commun. 1983, 256.
12. Rice, C. A.; Worley, S. D.; Curtis, C. W.; Guin, J. A.; Tarrer, A. R. J. Chem. Phys. 1981, 74, 6487; Worley, S. D.; Rice, C. A.; Mattson, G. A.; Curtis, C. W.; Guin, J. A.; Tarrer, A. R. ibid 1982, 76, 20, Worley, S. D.; Rice, C. A.; Mattson, G. A.; Curtis, C. W.; Guin, J. A.; Tarrer, A. R. J. Phys. Chem. 1982, 86, 2714.
13. See the many references quoted in reference 12.
14. Primet, M. J. Chem. Soc. Faraday Trans. 1 1978, 74, 2570; Primet, M.; Garbowski, E. Chem. Phys. Lett. 1980, 72, 472; Primet, M; Vedrine, J. C.; Naccache, C. J. Mol. Catal. 1978, 4, 411.
15. Van't Bilk, H. F. J.; Van Zon, J. B. A. D.; Huizinga, T.; Vis, J. C.; Koningsberger, D. C.; Prins, R. J. Phys. Chem. 1983, 87, 2264.
16. Solymosi, F.; Pasztor, M. J. Phys. Chem. 1985, 89, 4789.
17. See Uram, K. J.; Ng, L.; Folman, M.; Yates, J. T. J. Chem. Phys. 1986, 84, 2891 and the many references quoted therein.
18. Solymosi, F.; Tombacz, I.; Kocsis, M. J. Catal. 1982, 75, 78.

RECEIVED December 1, 1986

Chapter 11

Carbon Dioxide Reduction with an Electric Field Assisted Hydrogen Insertion Reaction

W. M. Ayers and M. Farley

Electron Transfer Technologies Inc., Princeton, NJ 08542

A method of chemically synthesizing reduced products including methanol from carbon dioxide and hydrogen has been developed. The method utilizes a metal hydride foil membrane as a continuous source of reactive surface hydrogen atoms and an electrostatic field to enhance the adsorption of carbon dioxide and bicarbonate onto the hydrogen rich surface. The subsequent chemical(rather than electrochemical) reaction between the adsorbed carbon dioxide and surface hydrogen/metal hydride results in the formation of reduced products.

Electrocatalysis at metal electrodes in aqueous ($\underline{1},\underline{2}$) and non-aqueous ($\underline{3}$) solvents ,phthalocyanine ($\underline{4}$) and ruthenium ($\underline{5}$) coated carbon, n-type semiconductors ($\underline{6},\underline{7},\underline{8}$),and photocathodes ($\underline{9},\underline{10}$) have been explored in an effort to develop effective catalysts for the synthesis of reduced products from carbon dioxide. The electrocatalytic and photocatalytic approaches have high faradaic efficiency of carbon dioxide reduction ($\underline{1},\underline{6}$), but very low current densities. Hence the rate of product formation is low. Increasing current densities to provide meaningful amounts of product, substantially reduces carbon dioxide reduction in favor of hydrogen evolution . This reduction in current efficiency is a difficult problem to surmount in light of the probable electrostatic repulsion of carbon dioxide, or the aqueous bicarbonate ion, from a negatively charged cathode ($\underline{11},\underline{12}$).

In an effort to overcome the limitations of electrochemical and photoelectrochemical cathodic reduction of carbon dioxide, and to draw upon insights of hydrogen insertion reactions studied in homogeneous

0097–6156/88/0363–0147$06.00/0

catalysis (13), we have developed a method that
electrostatically adsorbs carbon dioxide/bicarbonate at a
positively charged surface of a hydrogen atom
transmissive, metal hydride foil where the carbon dioxide
undergoes chemical reduction by the surface atomic
hydrogen/metal hydride (14).

Experimental Procedures

A modification of a bipolar membrane electrode
configuration, of the type used in diffusion studies, is
shown in Figure 1. On the left side of a metal hydride
foil, such as palladium, a constant current electrolysis
of a 1N sulfuric acid produces a high concentration of
hydrogen at that surface of the foil membrane. Some of
the hydrogen enters the palladium as atomic hydrogen. The
atomic hydrogen diffuses across the foil to the
opposite side due to the hydrogen concentration
gradient across the membrane (15).

On the right side of the membrane, a 0.1 M sodium
bicarbonate solution (saturated with carbon dioxide) is
in contact with the palladium. A potentiostat holds this
side of the membrane at a constant potential with
respect to a Ag/AgCl reference electrode. A platinum
counter electrode completes the circuit. The potential
on the carbon dioxide side of the membrane is held at a
positive potential with respect to the counter electrode
such that the adsorption of carbon dioxide as
bicarbonate anion can be enhanced by the electric field
between the metal hydride membrane and counter
electrode.

The overall reaction is electrochemical injection of
hydrogen into the left side of the membrane:

$$H+ \; + \; M + e- \; = \; MH$$

and reaction of the atomic hydrogen that diffuses to the
opposite side of the metal hydride foil to form reduced
products:

$$HCO_3^-/CO_2(ads) + 2MH \; = \; HCOOH \; + \; 2 \; M$$

$$CO_2 + 4MH \; = \; HCOH \; + \; H_2O$$

$$CO_2 + 6MH \; = \; CH_3OH \; + \; H_2O$$

where M/MH indicate the metal/metal hydride surface

Ideally, there should be no electrochemical oxidation
reactions on the carbon dioxide side of the membrane.
Hence, the power requirement would only be that required
to maintain a capacitive electrostatic field (no

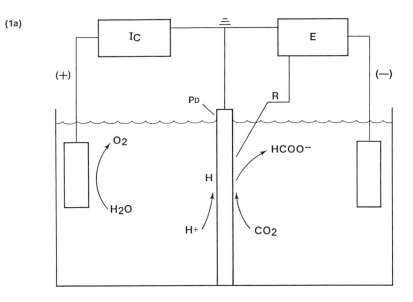

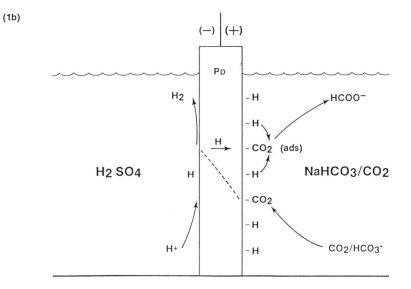

Figure 1. (a) Cell arrangement for metal hydride hydrogen insertion reaction. Left side, acid reduction and hydrogen atom incorporation in palladium (Pd) foil membrane. Right side, electrostatic field for enhancement of carbon dioxide/bicarbonate adsorption on foil membrane. (b.) Blow-up of palladium/hydride foil showing hydrogen insertion into carbon dioxide.

current). The polarization of the surface should be sufficient to adsorb the carbon dioxide/bicarbonate anion but less than that required to electrochemically oxidize the surface hydride to protons:

$$MH = M + H^+ + e$$

Therefore, a dielectric fluid with high carbon dioxide solubility and a non-reactive salt (e.g. NaCl) to provide a double layer, would be a better choice of fluid for this reaction. We find that in the electrolyte used in these experiments there is MH oxidation current as the membrane is biased to positive potentials above the metal hydride rest potential.

The metal hydride material in these experiments is a 25 micron thick palladium foil (99.999 purity, Alfa). The diffusion coefficient of hydrogen through palladium has been measured by electrochemical and gas phase techniques and is approximately 1.6e-7 cm^2/sec at 20 C (16). The exposed palladium membrane surface area is 2.29 cm^2. A platinum wire (area 0.08 cm^2) is the counter electrode for the capacitive field on the carbon dioxide side. A Nafion polymer membrane separates the counter electrode compartment from the palladium foil. The separator is not necessary for the process. However, it was utilized in these experiments as a precaution against diffusion into the reaction chamber of products that could theoretically be formed at the platinum wire counter electrode. Only trace amounts could theoretically be formed cathodically at the counter electrode since hydrogen evolution is favored over carbon dioxide on this metal (1), and the surface area is small (0.08 cm^2).

Results

The concentration of carbon dioxide reduction products (formic acid, formaldehyde, and methanol) in the reactor at the end of each run is listed in Table 1. The methanol concentration was determined by gas chromatography using a Porpac T column. Formic acid was determined with an Dionex ion chromatography column, and the formaldehyde was determined with a colorimetric chromatropic acid analysis (17). We did not examine the solution or gas phase for methane.

During each run, the membrane is electrochemically loaded with hydrogen from the left side with a constant electrolysis current of 30 mA. The initial potential of the palladium/palladium hydride (vs. Ag/AgCl reference electrode) on carbon dioxide reaction side at the start of the run is defined as E'. This potential depends on the hydride content of the membrane and the equilibrium between the metal hydride/bicarbonate solution.

The potentiostat on the carbon dioxide side is either
left in a floating mode (E') or set a some value
positive (more oxidizing) than E'. This potential is
defined as E". The difference between these two
potentials (E"-E') is called delta V in the table.
The reaction side solution is 0.1 M NaHCO3 saturated with
CO2 at 1 atm. Each run lasted 3 to 4 hours (specific
times in table).

The maximum rate of methanol formation is 3.9e-10
moles/cm^2-sec. This occurs at (E"-E') of 1.07 volts.
The formation rate corresponds to a turnover rate of 2.3
(assuming 10^{14} sites/cm^2).

The effect of the membrane potential difference,(E"-E'),
on the rate of methanol formation is shown in Figure 2.
As the potential is increased to more positive
potentials above the rest potential,E', of the
palladium/palladium hydride membrane,
oxidation of the surface hydrogen/palladium hydride to
protons, and oxidation of reduced carbon
dioxide intermediates can occur to reduce yields.
Hence there is some optimum
potential at which to hold the membrane potential to
maximize the rate of methanol formation. From this
preliminary data, it appears that the rate of methanol
formation decreases for E" more positive than 0.4 vs.
Ag/AgCl.

Discussion

We do not know from these esperiments if carbon dioxide
or bicarbonate is the more reactive species with the metal
hydride. The inference from the increase in rate with
increasing potential suggests that bicarbonate is
the intermediate. However, there is a chemical
equilibrium between carbon dioxide and bicarbonate. Hence
the rate dependence on positive potential may just
reflect increased surface coverage of bicarbonate
followed by equilibration with carbon dioxide.

It is interesting to consider a hydrogen mass balance in
the reaction. The ratio of the MH/H+
oxidation current (Table 1) to the hydrogen feed current
(30 MA) varies from no MH oxidation to 92% theoretical
oxidation (assuming none of the current is due to
reduced product oxidation).The maximum rate of methanol
formation (3.9e-10 moles/cm^2-sec) occured in a run that
lasted 10440 seconds. Hence the total hydrogen injected
into the membrane at 30mA would be 3.25 e-3 moles of
atomic hydrogen (assuming no loss from hydrogen
evolution). If all of this atomic hydrogen reacted
with the carbon dioxide on the opposite side of the
membrane,the CH$_3$OH concentration would be 962
ppm. Since the ratio of feed hydrogen to MH oxidation in

Table 1 Rate data for CH3OH   1.2M NaHCO3/CO2   pH 7.6

E' initial (volts)	E'' applied (volts)	delta V E''-E'	MH/H+ oxid. (mA)	duration (sec)	HCOOH (ppm)	HCOH (ppm)	CH3OH (ppm)	HCOOH rate	H2CO rate	CH3OH rate
										(gmoles/cm2-sec)
-0.41	0.80	1.30	27.5	12300	7.79	(.11	13.81	1.02E-10	(2.21e-12	2.60E-10
-0.58	0.60	1.23	15.5	14100	38.53	0.29	15.78	5.18E-10	1.37E-11	3.05E-10
-0.64	0.40	1.07	27.0	10440	(.1	(.11	15.71	(1.73e-12	(2.91e-12	3.90E-10
-0.20	0.20	0.66	13.5	21840	(.1	(.11	12.54	(8.69e-13	(1.46e-12	1.57E-10
-0.66	0.00	0.69	8.0	10980	11.07	0.12	12.06	2.39E-10	3.97E-12	3.74E-10
-0.47	-0.20	0.40	2.2	16920	20.37	0.13	11.01	2.28E-10	2.23E-12	1.77E-10
-0.41	-0.30	0.26	0.03	15120	27.87	0.12	8.77	3.50E-10	2.31E-12	1.58E-10
-0.20	-0.20	0.00	0.0	21360	15.65	0.12	6.18	1.24E-10	1.47E-12	7.10E-11
-0.50	-0.50	0.00	0.0	10020	21.22	(.11	4.29	4.42E-10	(3.51e-12	1.28E-10

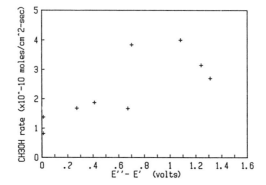

Figure 2. Methanol formation rate vs. positive bias
above   the rest potential, E'.

this experiment was 92%, in the worse case only 8% of the
surface hydrogen would be available for reaction with
the carbon dioxide. This would be equivalent to 77 ppm
CH_3OH. The measured concentration was 15.7 ppm.
It is not clear why the rate should increase up to
a certain limit of positive potentials even though the
availble surface hydrogen should be decreasing.

The thermodynamic potential of the palladium membrane
is a function of the hydrogen content in the membrane.
Palladium can dissolve approximately 60 atomic percent
hydrogen or $PdH_{0.6}$. Pourbaix approximates this as
Pd_2H and defines pH dependence of the potential as
($\underline{18}$):

$$E^o_{PdHx} = 0.048 - 0.059*pH$$

The carbon dioxide/bicarbonate reactions are ($\underline{18}$):

$$HCO_3^- + 7H^+ + 6e = CH_3OH + 2H_2O$$

$$E^o = 0.107 - 0.069pH$$

Hence, the Pd/PDH_x couple capable of directly reducing
carbon dioxide/bicarbonate to methanol.

These Nernstian equations should be modified to indicate
the effect of the electric field   superimposed
on the chemical potential of the surface hydrogen/metal
hydride. One possible way to treat this is with the
thermodynamic concept of electrochemical potential defined
by Guggenheim. The electrochemical potential is defined as:

$$E = E_o + e0_s$$

where E_o is controlled by the chemical constituents of
metal hydride/surface atomic hydrogen and the
electrostatic potential $e0$ is controlled by the
superimposed field. One interesting aspect of this
system is the ability to control both the chemical and
electrostatic components of the electrochemical
potential.

Summary

A new method of reducing carbon dioxide/bicarbonate
solutions to form methanol and other reduced products is
presented. The method overcomes the traditional
limitations of electrochemical reduction of carbon dioxide
at a cathode in that both the surface potential and
concentration of reducing species can be varied
independently.

Acknowledgment

This work was supported in part by Energy Conversion
Devices Inc.

Literature Cited

1. Kapusta,S.;Hackerman,N.J.Electrochem.Soc.
   1983,130,607-613.
2. Russell,P.G.;Kovac,N.;Srinivasan,S.;Steinberg,M.
   J.Electrochem. Soc1977,124,1329-1338.
3. Roberts,J.L.;Sawyer,D.T. J.Electroanal. Chem.
   1965,9,1.
4. Kapusta,S;Hackerman,N.J. Electrochem. Soc.
   1984,131,1511-1514.
5. Frese,K.W.;Leach,S.J.Electrochem. Soc.
   1985,132,259-260.
6. Frese,K.W.;Canfield,D.J.Electrochem. Soc.
   1984,131,2518-2522.
7. Sears,W.M.;Morrison,S.R.J.Phys. Chem.
   1985,89,3295-3298.
8. Monnier,A; Augustynski,J; Stalder,CJ. Electroanal.
   Chem.,1980,112,383-385.
9. Inoue,T; Fujishima,A;Kanishi,S.;Honda,K. Nature,
   1979,277,637
10. Aurian-Blanjeni,B.;Taniguchi,I.;Bockris,J.O'M.
    J.Electronal. Chem.,1983,149,291-293.
11. Babenko,S.D.;Benderskii,V.A.;Krivenko,A.G.;Kurmaz,V.A.
    J.Electroanal. Chem. 1983,159,163-181.
12. Hori,Y.;Suzuki,S.J.Electrochem.Soc.1983,130,2387-
    2390.
13. Eisenberg,R.;Hendriksen,D. In Advances in
    Catalysis;Eloy,D,Pines,H.; Weisz,P.,Eds.; Academic
    Press:New York,1979; Vol. 28, p. 79.
14. Ayers,W. U.S.Patent 4 547 273,1985.(filed June
    7,1984).
15. Subramanyn,P.K. In Comprehensive Treatise of
    ElectrochemistryBockris,J.O'M.,Ed.;Plenum Press:New
    York,1981;Vol.4,411-462.
16. Fukai,Y.;Sugimoto,H.Adv.in Physics 1985,34,269.
17. Analysis preformed by National Spectrographic
    Laboratories, Cleveland, Ohio.
18. Pourbaix,M Atlas of Electrochemical Equilibria;
    Pergamon Press:New York,1966;p.358.

RECEIVED September 25, 1987

Chapter 12

Electrochemical Reduction of Aqueous Carbon Dioxide at Electroplated Ru Electrodes

Investigations Toward the Mechanism of Methane Formation

Karl W. Frese, Jr., and David P. Summers

SRI International, Menlo Park, CA 94025

The effect of experimental conditions on methane formation was studied to gain information on the rate determining step. Cyclic voltammetry performed in the presence and absence of CO_2 suggested the presence of carbon containing intermediates that block surface hydrogen evolution sites. The maximum in the CH_4 formation rate with pH implies that the rate increases with increasing hydrogen coverage on the electrode until the coverage becomes so high that sites for CH_4 formation are blocked. The effect of potential on CH_4 formation indicated that CH_4 evolution occurs at potentials at which an appreciable hydride coverage exists, also indicating the importance of surface hydrides. The rate of CH_4 formation increases with temperature, but at $T>85°C$ the electrode becomes deactivated because of a surface carbon species. An activation energy for CH_4 formation of ~9 kcal mol^{-1} is inferred. Electrolyte impurities are implicated as promoters in the formation of CH_4 in reagent grade sodium sulfate.

The goals of replacing finite world natural gas reserves and producing fuels from inorganic sources and solar energy has been a motivating force for studying the electrochemical reduction of CO_2 to CH_4. Although initial work focused on semiconductor electrodes in order to capitalize on their potential ability to directly utilize light energy such efforts have only lead to the formation of methanol (1-5). In order to improve the catalytic properties of our electrodes we turned to metal electrodes, which can be coupled to a photovoltaic cell. We had previously reported that CH_4 could be formed from the reduction of CO_2 in aqueous solution at Ru electrodes (6). Results presented here substantially extend those results and provide detailed information on the effect of pH, electrode potential, temperature, electrolysis time, and electrolyte purity.

0097-6156/88/0363-0155$06.00/0
© 1988 American Chemical Society

<u>EXPERIMENTAL</u>

Electrolytes were CO_2-saturated (1 atm) aqueous solutions of either 0.2 M reagent grade sodium sulfate, 0.2 M 99.999% sodium sulfate or 0.05 M reagent grade sulfuric acid in distilled deionized water (Millipore).

Electrodes were prepared by plating Ru metal onto rods of spectroscopic carbon as previously described (6). The geometrical area of the electrodes was 3 cm^2 ± 20 %. Each entry in the tables and figures was obtained on different days with the electrode kept in ordinary laboratory air overnight between runs.

A two compartment cell was employed to avoid oxidation of the CO_2 reduction products. This procedure allowed pH changes of 1-2 units to occur during the electrolysis.

Analysis was by gas chromatography, full analytical procedures and Auger spectroscopy are described elsewhere (Summers, D. P. and Frese Jr., K. W. Submitted to <u>J.Amer. Chem. Soc.</u>). Samples for Auger analysis were removed under potentiostatic control, rinsed with millipore water, and allowed to dry before mounting on sample holder.

<u>RESULTS</u>

<u>Scanning Electron Microscopy</u>. This paper focuses on results obtained with electroplated Ru electrodes similar to those used previously to reduce CO_2 to methane (6). Typical Scanning Electron Micrographs (SEM's) of such an electroplated Ru electrode are shown in Figure 1. The surface shown is identical to that of all other electrodes investigated regardless of whether they are fresh from the plating bath or have been used in many different electrolysis experiments. The surface consists of fused spheroids (~3μm) of Ru formed around sites for nucleation during plating as is commonly found for electrodeposited surfaces. In Figure 1b cracks that are present on the surface can be seen completely splitting many of the Ru spheroids indicating that they are formed after most, if not all, the Ru has been deposited. Internally stressed electroplates are common in the electrodeposition of metals (7) and are the probable source of the cracking. A SEM of a section of Ru plate that lifted up during sample preparation showed the thickness of the plates to be ~4 μm.

<u>Inactivity of carbon surfaces</u>. Since the Ru plate is cracked, probably allowing electrolyte access to the carbon substrate, the question of formation of CH_4 at the carbon arises. Controlled current electrolysis (100 μA cm^{-2}) using a bare carbon rod, similar to that used as a substrate for Ru plating, was performed in a CO_2 saturated reagent 0.2 M Na_2SO_4 solution at pH 4 and 57°C. Although methanol was detected in significant yields (70%), no CH_4 was detected even at such negative electrolysis potentials, -1.0 to -1.2 V vs SCE. The cyclic voltammetry (Figure 2a) of a carbon rod in 0.2 M Na_2SO_4 at 60°C in the presence and absence of CO_2 shows no excess current for CO_2 reduction until ~-0.8 V vs SCE. The excess current that occurs at E > -0.8 V vs SCE correlates with a significant rate of methanol production. Under identical conditions, CH_4 is formed at Ru for electrode potentials cathodic of -0.48V SCE. Finally, it was

A

B

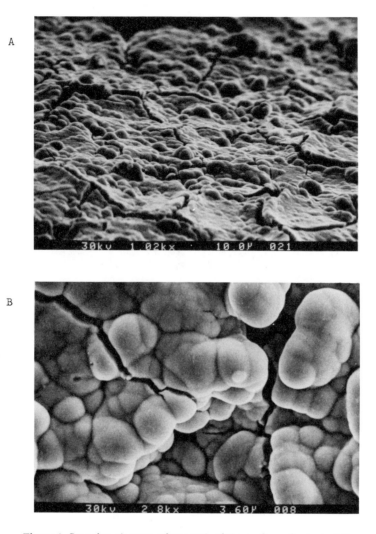

Figure 1. Scanning electron micrograph of the surface of a typical Ru plate electrode used for aqueous CO_2 reduction at magnifications of A, 800 × and B, 2200 ×. Picture A was taken at an angle to enhance perspective; picture B was taken straight on.

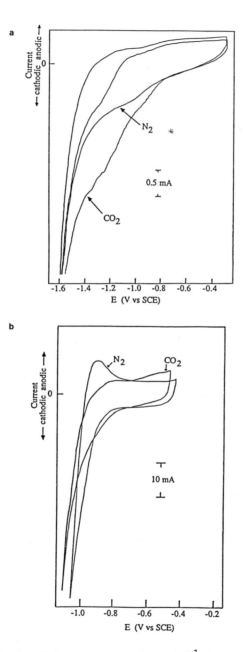

Figure 2. Cyclic Voltammetry at 3 V min^{-1} in a quiet solution saturated with N_2 or CO_2 of a) a carbon rod electrode or b) a Ru electroplated electrode.

shown previously that CH_4 could be formed at electrodes made from teflon bonded Ru sponge as well as from electroplated electrodes (6) indicating that CH_4 does not arise from the hydrogenation of carbon atoms from the carbon substrate. Thus we can rule out any activity of the carbon rod itself towards CH_4 formation.

Current/Voltage Curves. Figure 2b shows the cyclic voltammetry of an electroplated Ru electrode at 60°C in the presence and absence of CO_2 in a pH 4.2, 0.2 M Na_2SO_4 solution. The first feature of note is the absence, in a CO_2 saturated solution, of the anodic peak at -0.9 V due to hydrogen oxidation that is present in a CO_2 free solution. The second feature is that, as the potential moves into the hydrogen evolution region, currents rise much more slowly in a CO_2 saturated solution compared to a nitrogen saturated solution. Both these observations are consistent with a model in which sites for hydrogen evolution are blocked by CO_2 reduction intermediates or products. The loss of the hydrogen oxidation peak suggests either that the CO_2 reacts with the hydrogen or that carbonaceous species are formed from CO_2 that block sites for hydrogen evolution.

Methane Formation Rate. A plot of the average rate of CH_4 formation vs total electrolysis time for many experiments at 60°C, initial pH 4 in both reagent grade and 99.999% Na_2SO_4 is shown in Figure 3. The data for a Ru electrode in pure 0.05 M H_2SO_4 or 0.05 M H_2SO_4/0.2M reagent Na_2SO_4 is very similar to that for experiments conducted at an initial pH of 4 with respect to the decline in rate and electrolyte purity.

An effect of electrolyte purity is also evident, as shown by the 5-6 times lower rate in the high purity electrolyte. The effect appears to be due to the electrodeposition of adventitious impurities. Electrochemical stripping experiments performed on a solution made from reagent grade Na_2SO_4 revealed the presence of Zn, Cu, and As and Auger spectroscopy of electrodes used to electrolyze such solutions showed Cu, Fe, Ni, and Zn on the surface. An electrode that was used in several CO_2 electrolysis experiments showed surface Cu by Auger spectroscopy (see below). We will report on these results more fully along with results of experiments in progress in a future paper.

It is seen that the CH_4 formation rate declines with time in the closed system. Although this behavior could represent deactivation of the electrode surface by the formation of carbon containing species (either blocking intermediates or side products such as CO, see below), the effect may be caused by pH changes that occur as protons are consumed during the reaction. A plot of the total current during the electrolysis of a pH 4, 0.2 M Na_2SO_4, CO_2 saturated solution at 60°C shows that the decline in rate with time is mirrored by a decline in the total current (which is partially due to hydrogen evolution, Summers, D. P. and Frese Jr., K. W. Submitted to J.Amer. Chem. Soc.). However, the same decline is also seen when a CO_2 free solution is electrolyzed. Also, if we electrolyze a pH 4, 0.2 M Na_2SO_4 nitrogen-saturated electrolyte before and after electrolysis of a CO_2 solution, a nearly identical decline in current with time is seen regardless of the exposure to CO_2. Thus the drop in current is not due to irreversible

Table 1
Activity for Methane Formation of a Ru Electrode Over Multiple
Electrolyses.[a]

pH[b]/run	j[c] (μA cm^{-2})	Rate[d] (mol cm^{-2} hr^{-1})	Eff.[e] (%)
T = 60°C			
4/1	160	4.3 x 10^{-8}	5.7
4/2	140	3.1 x 10^{-8}	4.7
4/3	88	3.3 x 10^{-8}	8.0
4/4	87	4.1 x 10^{-8}	10.0
2.9/1	366	14.3 x 10^{-8}	12.2
2.9/2	328	15.9 x 10^{-8}	10.4
2.9/3	239	14.3 x 10^{-8}	12.7
T = 80°C			
2.7/1	392	46 x 10^{-8}	25
2.7/2	289	36 x 10^{-8}	27
2.7/3	224	24 x 10^{-8}	23

a) All electrolysis times are 5-6 hrs in 0.2 M reagent
   grade Na_2SO_4 at 60°C and -0.545 V vs SCE with an
   initial pH of 4.
b) Initial pH.
c) Average current density based on geometrical area.
d) Average rate of methane formation.
e) Faradaic efficiency for methane formation.

deactivation of the electrode by CO_2 or its reduction products or
intermediates.
    In Table 1 data on the effect of multiple electrolyses on the
CH_4 formation rate is presented. It is apparent from the data at
60°C that if an electrode that has been used to electrolyze a CO_2
saturated solution is reused no measurable decrease in average CH_4
formation rate is observed. Thus, at 60°C, any drop in rate that
occurs during an individual experiment is not due to an irreversible
deactivation of the catalytic surface.   A different conclusion is
reached after electrolysis at t> 80°C (see below).

Effect of Added CO. Carbon monoxide is formed in CO_2 reduction
experiments with faradaic efficiencies typically 1 %, but a range of
) to 10% has been observed. It is important to consider the effect
of CO on the methane formation rate because CO is known to be a
strongly chemisorbing precursor to methane. The effect of added CO
at levels that are 100-1000 times higher than normally found in CO_2
electrolysis experiments experiments is to lower the average
current, CH_4 formation rate, and faradaic efficiency. Usually lower
current increases the faradaic efficiency with only a small effect
on the CH_4 formation rate. Carbon monoxide does indeed inhibit the
the formation of CH_4, but the effect is not very great at the CO
concentrations employed, the rate of CH_4 formation being reduced by
about a factor of two. The inhibition effect likely occurs by the
blocking of sites for CO_2/H^+ reduction by the more strongly bound
CO.

Effect of Added Hydrogen. Added hydrogen gas also has an effect on
the rate of CH_4 formation. If the partial pressure of H_2 is
increased by a factor of ~2000 or ~4200 above a typical value of 3 x
10^{-6} atm, the CH_4 formation rate increases by 10% and 41%
respectively. These results demonstrate that hydrogen atom coverage
is an important factor in the rate of CH_4 formation. The relatively
small increase in CH_4 formation rate for such a large increase in H_2
partial pressure could mean that the solution next to the surface is
near saturation with H_2. However, the shape of the hydrogen
adsorption isotherm is not known and it is possible that the
hydrogen coverage on the surface may not vary greatly over the
concentrations of hydrogen gas used.

Effect of pH. The effect of the pH on the rate of CH_4 formation for
two electrodes is shown at the top of Figure 4 . The data indicate
that CH_4 can be made at pH values as alkaline as 9.1 with modest
rate and faradaic efficiencies. This might indicate that direct
reduction of bicarbonate ions occurs but even at alkaline pH a
significant partial pressure of CO_2 is present (at all other pH's
solutions were made from gaseous CO_2 and had a constant CO_2 partial
pressure of 1 atm, see experimental). Indeed, analysis of the gas
over the solution indicated the presence of more CO_2 than our gas
chromatograph could measure (~0.1 atm). However CH_4 can be formed at
pH values below 1 where there is not a significant concentration of
bicarbonate ions. Therefore, at acidic pH's at least, reduction of
CO_2 or H_2CO_3 and not HCO_3^- occurs. However the concentration of
carbonic acid at a CO_2 partial pressure of 1 atm is probably much
too low (8) (6 x 10^{-5} M) to support the observed CH_4 formation rates
even assuming diffusion control.

As Figure 4 shows, the CH_4 formation rate does depend on pH. In
the pH region 9 to ~3 the rate increases. This effect is
rationalized as occurring either because of an increased surface
hydride coverage increasing the rate of hydrogenation of CO_2
reduction intermediates or an increased rate of oxygen removal from
the surface favoring the deoxygenation of CO_2 or its intermediates.
At pH's less than 2-3 the rate begins to decrease. This may be due
to a coverage of surface hydrides that is so high that sites for CO_2
reduction are blocked.

Effect of Electrode Potential. Figure 5 and Table 2 summarize the
results concerning the influence of electrode potential on the CH_4
formation rate in a CO_2 saturated, 0.2 M Na_2SO_4 solution at 60°C.
The data shows an apparent linear dependence of the CH_4 formation
rate on potential. The increase in rate with potential is not
unexpected since the CH_4 formation reaction is too slow to be
diffusion controlled. The rate drops to zero at -0.48 V vs SCE. With
an average pH of ~5 and a partial pressure of hydrogen estimated to
be 1 x 10^{-6} atm, the reversible potential for hydrogen evolution is
-0.36 V vs SCE. The potential at which the CH_4 formation begins is
~120 mV cathodic of the formal hydrogen potential and hence the
electrode probably has an appreciable hydrogen coverage at
potentials were CH_4 is formed. This is consistent with a model for
the reduction of CO_2 in which surface carbon intermediates are
hydrogenated with surface hydrides in a key step. At more cathodic

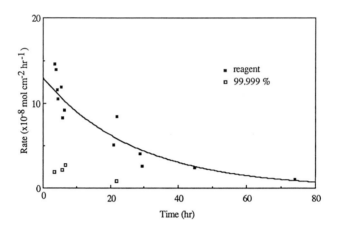

Figure 3. Average methane formation rate versus total electrolysis time (closed symbols = reagent grade Na_2SO_4; open symbols = 99.999% Na_2SO_4).

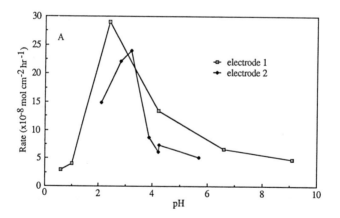

Figure 4. Plot of the rate of methane formation at a Ru electrode as a function of pH at 60-63°C and a constant overpotential.

Table 2
Effect of potential on the rate of methane formation at Ru
electrodes.[a]

V (V vs SCE)	Q^b (coul)	j^c (μA cm^{-2})	Rate[d] (mol cm^{-2} hr^{-1})	Eff[e] (%)
-0.48	1.3	22	0	0
-0.50	2.0	30	1.6×10^{-8}	10.8
-0.51	3.4	57	2.8×10^{-8}	10.2
-0.545	3.3	60	4.3×10^{-8}	15.4
-0.545	4.8	86	4.1×10^{-8}	10.3
-0.60	5.8	102	6.8×10^{-8}	14.2

a) In 0.2M Na_2SO_4 at 60°C and an initial pH of 4. All electrolysis
   times were ~5 hrs.
b) Total charge passed.
c) Average current density based on geometrical area.
d) Average rate of methane formation.
e) Faradaic efficiency for methane formation.

potentials the hydrogen coverage increases, thereby increasing the
rate.

Effect of Temperature. The effect of temperature on the rate of CH_4
formation at a single Ru electrode at -0.545 V vs SCE is shown in
Figure 6; the data are listed in Table 3. The inset illustrates a
plot of faradaic efficiency vs temperature. It is note worthy that
an efficiency of 42% at 80°C is the highest recorded for CH_4
formation. The increase in the faradaic efficiency for CH_4 formation
indicates that CH_4 formation increases faster with temperature than

Table 3
Effect of temperature on methane formation rate.[a]

Run[b]	T (°C)	Q^c (coul)	j^d (μA cm^{-2})	Rate[e] (mol cm^{-2} hr^{-1})	Eff (%)
1	41	4.6	78	1.2×10^{-8}	3.4
4	50	2.8	51	2.9×10^{-8}	12.1
7	60	2.5	45	1.6×10^{-8}	7.8
2	61	3.0	68	5.9×10^{-8}	18
3	71	3.0	58	8.4×10^{-8}	31
5	82	2.4	43	8.4×10^{-8}	42
6	90	2.4	44	3.9×10^{-8}	19

a) In 0.2M Na_2SO_4 at -0.545 V and an initial pH of 4. All
   electrolysis times were 5-6 hrs using the same electrode.
b) Order of experiment.
c) Total charge passed.
d) Average current density based on geometrical area.
e) Average rate of methane formation.

competing reactions, e.g. H_2 evolution, and so must have a higher activation energy.

The temperature dependence experiments are numbered in the order in which they were conducted. For the first four experiments (each done on successive days at t<75 °C), the order of the experiments did not affect the performance of the electrode. However, if the electrode is used in electrolyses above ~85°C, it begins to deactivate. The experiment at 90°C led to a decline in the CH_4 rate and, when the electrode was used at 60°C after the 90°C experiment, the CH_4 formation rate was significantly reduced (as was the faradaic efficiency). Thus at t>~80°C, an irreversible deactivation does occur.

This is confirmed by the data on multiple electrolyses (Table 1). It can be seen that if an electrode is used several times to electrolyze a CO_2 saturated solution at 60°C no decline in average CH_4 formation rate is observed, but at 80°C a clear decline in CH_4 formation is seen from one experiment to the next. Such behavior is consistent with a slow deactivation of the electrode surface at higher (>~80°C) temperatures.

An Arrhenius plot using the low temperature data (last three points), Fig. 6, gives an activation energy of ~9 kcal mol^{-1}. McKee (9) observed an activation energy of 9.1 kcal mol^{-1} for the rate of formation of CH_4 from H_2 and CO on unsupported Ru catalysts in the temperature range from 25°C to 150°C. McKee also observed a curvature in his Arrhenius plot similar to that seen in Figure 6 although at a slightly higher temperature.

Auger Electron Spectroscopy. After the last 60°C experiment, No.7 in table 3 and Figure 6, the Auger electron spectrum of the Ru surface was determined as shown in Figure 7a. Compared to the Auger spectrum (10) of a clean Ru surface, the largest peak due to Ru is changed in symmetry and size. The signal is not highly symmetrical with only one positive peak for the highest energy signal, but highly unsymmetrical and an exhibits two positive peaks. By comparison, the Auger spectrum of the used Ru surface after Ar$^+$ sputtering (Figure 7b) shows the same Ru signal as that of a carbon free Ru surface exhibiting a highly symmetrical signal with only one positive peak. These characteristics are indicative of the presence of large amount of carbon on the surface due to the accidental overlap of the Auger signal due to carbon (10) Since the carbon signal disappears when the electrode is Ar$^+$ sputtered the signal is due to carbon on the surface of the Ru and not due to the carbon substrate. Thus it appears that the deactivation is caused by the formation of surface carbonaceous species at high temperatures. This does not occur at lower temperatures since an electrode used once at 60°C does not show the presence of large amounts of surface carbon. Sputtering also lessens the Cu impurity signal (Figure 7a) proving it is a surface species.

It has been postulated (11-12) that the formation of CH_4 by the heterogeneous catalytic reduction of CO gas with gaseous hydrogen proceeds via carbon atoms on the surface from CO. Such a mechanism involving dissociative adsorption of CO may operate during the electrochemical reduction of CO_2 in aqueous solution. This leads us to a tentative conclusion that the deactivation of the electrode occurs because of polymerization of surface carbon atoms to an

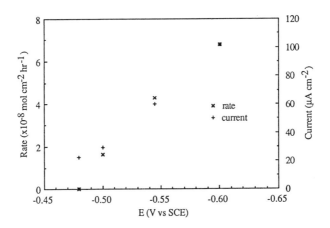

Figure 5. Plot of methane formation rate and current vs potential for electrochemical reduction of CO_2 at Ru electrodes. Hydrogen couple formal potential, -0.36 V vs SCE.

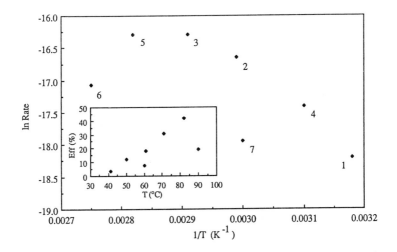

Figure 6. Plot of ln of methane formation rate vs temperature and faradaic efficiency vs temperature for electrochemical reduction of CO_2 at Ru electrodes.

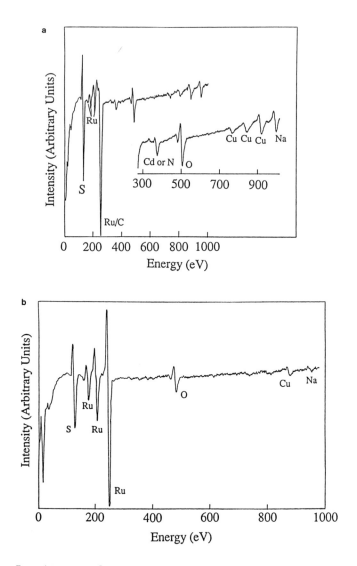

Figure 7. a) Auger electron spectrum of the surface of a Ru electrode after repeated electrolysis of a CO_2 saturated aqueous solution at 40 to 90°C (see text). b) Auger electron spectrum of the surface after Ar+ sputtering of the Ru electrode used in a).

inactive form of carbon. This may occur because of a depletion of hydrogen atoms on the surface, an excess coverage of carbon at higher temperatures or it may be that at higher temperature carbon atoms are more mobile and can move across the surface to combine.

Discussion

A case has been (13) made for a mechanistic commonality between gas/solid and electrocatalytic approaches to similar reactions such as the interaction of hydrogen molecules or CO with Pt surfaces. Unsupported Ru has exceptional activity for methanation and Fischer-Tropsch type gas/solid reactions (12). The electrochemical formation of CH_4 has only been observed on Ru and not with other materials such as Pt, Mo, C, Pd, Ag, Os, Ni, GaAs, GaP, and Si (14) Evidently the exceptional character of Ru in gas phase reactions is carried over in electrochemical systems. It is useful to discuss our electrochemical results vis à vis what is known about the gas/solid methanation reaction. However the formation of CH_4 from CO (15) rather than CO_2 (16), is much better characterized.

Consider what is clear. The pH dependence of the CH_4 rate at constant overpotential has a pronounced maximum. A rate limiting surface process involving H atoms is suggested. That CH_4 is only formed at Ru electrodes indicates the importance of a surface catalytic process. There is strong support in the literature (11-12) for the existence of surface carbon atoms formed from CO dissociation and that hydrogenation of the active form leads to CH_4. Although rate limitation by chemical dissociation of CO or CO_2 is plausible, the maximum in the pH dependence would seem to rule out such a limitation. The coverage with hydrogen atoms would decline for any increase in pH and thus more free sites for oxygen or carbon would be present and the CH_4 rate would increase without a maximum, contrary to observation. The CO_2 reduction current leading to methanol on GaAs also has a similar maximum (4) and it was concluded that the rate limiting step is a chemical combination of a surface H atom and a carbon-containing intermediate. Our results here support such a conclusion.

The CH_4 rate did not saturate in the electrode potential range investigated. It may be concluded that the surface is not saturated with intermediates at pH 4-5 at 60°C. The enhancement in the rate upon addition of hydrogen gas is consistent with a) an unsaturated surface, and b) the increase in CH_4 rate for pH < 4. Both adding hydrogen and lowering the pH with respect to normal pH 4 conditions leads to higher hydrogen coverage and higher CH_4 rate. More cathodic potentials also lead to higher rates and the probable cause is increased hydrogen coverage. These observations are consistent with a rate limiting step involving a surface hydrogenation.

Our conclusion from the Auger results are in agreement with known aspects of the formation of CH_4 from gaseous CO on Ru surfaces. When the hydrogen coverage is relatively low and the temperature relatively high, carbon builds up on the Ru surface leading to partial deactivation. Similarly, excess carbon build up is seen for low H_2/CO ratios and high temperatures in the gas phase. A buildup of surface carbon in our system also does not support rate limitation by dissociation of CO or CO_2.

All our results favor a mechanism with hydrogenation of a surface carbon species as the rate limiting step (eq. 1). The low

$$H_{ad} + CH_x \rightarrow CH_{x+1} \qquad (1)$$

where x = 0 to 3

CH_4 rate at very low pH suggests that undissociated CO is not a major adsorbed species. If CO dissociation was necessary for the formation of surface carbon for CH_4 formation, then it is very difficult to accept the lower CH_4 rate at low pH because it is unlikely that a hydrogen would displace a chemisorbed CO molecule. The effect of low pH may be due to blocking the formation of an intermediate more weakly bound than hydrogen, perhaps formate radicals.

Further insight into the energetics of the possible elementary steps leading to methane is obtained using model calculations of binding energies of chemisorbed species obtained by means of the Polar Covalence Model (17, Karl W. Frese, Jr. Surf. Sci., in press). Table 4 shows a set of consecutive hydrogenation reactions using

Table 4.
Model Calculations of Energetics of Possible Elementary Steps in Methane Formation on Ru Surfaces.

Step	ΔH^a (kcal mol^{-1})	ΔH^b (kcal mole^{-1})
1. Ru_2-C + H = Ru_2-CH	- 26	-46
2. Ru_2-CH + H = $Ru=CH_2$	+ 9	-11
3. $Ru=CH_2$ + H = $Ru-CH_3$	+ 7	-13
4. $Ru-CH_3$ + H = $Ru + CH_4$	+ 43	+23[c]

a) at zero coverage.
b) at 0.5 H_{ad} coverage
c) Calculated Gibbs energy for step 4 is considerably more exothermic, see text.

chemisorbed carbon and hydrogen atoms as initial reactants. This mechanism (11-12,18) is thought to be part of the most likely pathway to methane from CO.

The calculated results suggest the most difficult step may be hydrogenation of the Ru-methyl complex on the basis of unfavorable energetics. Because hydrogen atoms are postulated as reactants in the rate limiting step in methane formation and because of competetive hydrogen evolution, the steady state coverage of H_{ad}, although unknown, has to be significant. Assumimg 0.5 coverage, it follows that the heat of each of the steps would be more favorable by about 20 kcal mol^{-1} because of the known decrease in the heat of chemisorption (19) of H atoms with coverage on Ru. Furthermore, a very significant entropy effect amounting to approximately 26 to 45 cal mol^{-1} K^{-1}, depending on whether hydrogen is described as a two dimensional gas or an immobile species, would cause the Gibbs energy of last step (4) to be as low as +15 to +9 kcal mol^{-1} a 80°C. (These energy estimates assume the entropy changes involving the adsorbed carbonaceous intermediates are small enough to be ignored.) The coverage and entropy effects do not alter the conclusion that step (4) is the least favorable energetically and the most likely rate determining step in qualitative agreement with our discussion above, eq.1. It is of interest to note that step 4 has been recently ruled out (18) as rate limiting in the formation of methane from CO on Ni according to the pathway in Table 4. Model calculations (Karl W. Frese, Jr. *Surf. Sci.*, in press) similar those described for Ru above suggest that in the case of Ni, the Gibbs energy change for step (4) with half hydrogen coverage is ~25 kcal mol^{-1} lower (ΔG = -10 to -15 kcal mol^{-1}).

These results should be viewed as a survey of the important factors affecting the rate of CH$_4$ formation and not a detailed kinetic study in which very strong support can be given for a particular mechanism and rate determining step. Given the complexity of the system and the apparent role of impurities, much more effort will be needed to discern the most likely steps. Nevertheless the data presented do provide important clues as to important surface reactions and possible rate determining steps. Future work will help to identify the rate determining step more clearly and will clarify the role of impurities in enhancing the catalytic activity of Ru.

Acknowledgments

The authors wish to gratefully acknowledge the support of the Gas Research Institute.

Literature Cited

1. Halmann, M. Nature 1978,275,115.
2. Inoue, T., Fujishima, A., and Hounda, K. Nature 1979,277,637.
3. Canfield, D. and Frese Jr., K. W. J. Electrochem. Soc. 1983,130,1772.
4. Frese Jr., K. W. and Canfield, D. J. Electrochem. Soc. 1984,131,2614.
5. Summers D. P.,Leach S., and Frese Jr., K. W. J. Electroanal. Chem. 1986,205,219.
6. Frese Jr., K. W. and Leach, S. J. Electrochem. Soc. 1985,132,259.
7. For example see; Pearlstein, F. and Weightman, R. F. Plating 1969,56.1158.
8. Kern D. J. Chem. Ed. 1960,37,14.
9. McKee D. J. Catalysis 1967,8,240.
10. Goodman, D. and White J. Surface Science 1972,90,201 and references therein.
11. Wise H. and McCarty J. Surface Science 1983,133,311.
12. Winslow P. and Bell A. J. Catalysis 1985,94,385.
13. Stonehart, P. and Ross, P.N. Catal. Rev.-Sci. Eng. 1975,12,1.

14. SRI International Annual Reports for the Gas Research Institute
    (1984-86), SRI Proj. No. PYH 7142, GRI Contract No.
    5083-260-0922.
15. Solymosi, F., Erdohelyi, A., and Kocsis, M. J. Chem. Soc.,
    Faraday Trans. 1981,77,1003.
16. Vannice, M. A. J. Catalysis 1975,37,462.
17. R.T. Sanderson, Polar Covalence, Academic Press, New York
    (1983).
18. J.T. Yates, S.M. Gates, and J.N. Russell, Jr., Surf. Sci.
    1985,164, L839.
19. K. Kraemer, and D. Menzel, Ber. Bunseges. Gesell. 1974,78,728.

RECEIVED December 1, 1986

Chapter 13

Electrochemical Studies of Carbon Dioxide and Sodium Formate in Aqueous Solutions

M. H. Miles and **A. N. Fletcher**

Chemistry Division, Research Department, Naval Weapons Center, China Lake, CA 93555

Protonated species such as H_3O^+ must be minimized in order to obtain favorable conditions for the electrochemical reduction of carbon dioxide or sodium formate in aqueous solutions. Nearly neutral electrolytes that do not function as effective proton donors or acceptors provide an extra 0.3 V potential window for cathodic reactions between the RHE potential and the actual reduction of water molecules. Investigations of NaCOOH on 30 different electrode materials show that the electrochemical reactions of formate are controlled mainly by the adsorption of HCO_2^- on the electrode surface. Appreciable formate adsorption occurs only for Rh, Pd, Ir, Pt, and Au. Evidence for CO_2 reduction was observed on precious metals and their alloys such as Pt, Ir, Pd, Pt-Ru alloy and Ru-Ir alloy. The catalytic activity found for precious metals in nearly neutral electrolytes may reflect their favorable adsorption of both carbon dioxide molecules and formate ions.

The conversion of carbon dioxide and water into methanol and oxygen

$$CO_2 + 2\ H_2O \rightarrow CH_3OH + 3/2\ O_2\uparrow \qquad (1)$$

requires an input of energy ($\Delta H° = 719.238$ kJ/mol). However, the thermodynamic potential for this reaction ($E° = -1.976$ V) is actually less than that for the electrolysis of water ($E° = -1.2288$ V). The major source of interest in this reaction is as a means of converting carbon dioxide into organic compounds and portable fuels (1-3). The electrochemical reduction of CO_2 in aqueous solutions using metal electrodes generally yields formic acid and formate ions as the main products (2, 3). The major experimental difficulty in the cathodic reaction is the further reduction of formic acid to methanol (3, 4). Previous studies of the formic acid reduction step in acidic solutions have shown that H_3O^+ rather than HCOOH is reduced (4). Undissociated HCOOH molecules aid the reduction of

H_3O^+ by serving as a conveyor of protons to the electrode surface. This study is focused on neutral or alkaline solutions in order to minimize the reduction of H_3O^+.

Experimental

All electrochemical measurements were performed using a beaker-type cell where the platinum counter electrode was isolated from the main compartment by a section of glass tubing with an ultra-fine frit at the bottom. Most electrodes were prepared from metals (Alfa) with purity generally better than 99.9%. The 50-50 atomic percent alloys of Ru-Ir and Pt-Ru were prepared by International Nickel Company. The ruthenium metal tested was also provided by this company. Metallic wires were sealed in glass using shrink Teflon tubing. When this was not possible, silicon rubber (RTV 102) or epoxy (Epoxi-Patch) was used. The supporting electrolyte solutions were prepared with reagent grade salts and distilled, de-ionized water. All potentials were measured against a saturated calomel electrode (SCE). Electrochemical studies of NaCOOH (G. F. Smith, Reagent) were always made in helium-saturated solutions. Saturated carbon dioxide solutions were prepared by bubbling the cylinder gas (Matheson, 99.99% min, <20 ppm oxygen) through the aqueous electrolyte for several minutes.

The palladium-hydrogen (Pd-H) electrode used to monitor the solution pH was prepared each day by the procedure described by Gileadi (5). The relationship between pH and potential (E) versus SCE is given by

$$pH = -(E + 0.2031)/0.0575 \qquad (2)$$

at 23°C and ambient pressure (93 kPa). The Pd-H electrode served another important function by providing a measure of the reversible hydrogen electrode (RHE) potential under various conditions. The Pd-H electrode gives a constant potential of approximately +0.050 V versus RHE (5).

Electrochemical experiments involved the use of a potentiostat/galvanostat, a current-to-voltage converter, and a universal programmer (PAR Models 173, 176 or 179, and 175). Results were displayed using either an X-Y recorder (Hewlett-Packard 7047A) or a digital oscilloscope (Nicolet 4094 A). The Pd-H versus SCE potentials were measured with both the PAR potentiostat and a digital multimeter (Keithley 175).

Results and Discussion

Cyclic voltammetric traces obtained with a molybdenum electrode in 3.0 m $Mg(ClO_4)_2$ at pH = 8.0 are shown in Figure 1. The positive-going potential scan is limited by the oxidation of the molybdenum electrode. The addition of 0.50 m NaCOOH produces a new cathodic peak near -0.8 V with this electrode. However, the peak current corrected for the background current (dashed line) is only 0.035 mA (0.18 mA/cm^2). Theoretical peak current densities for irreversible electrode reactions that are diffusion controlled are given by

$$i_p = 3.01 \times 10^5 n(\alpha n_a)^{1/2} D^{1/2} C \nu^{1/2} \tag{3}$$

For the experimental conditions used ($\nu = 0.020$ V/s, $C = 5 \times 10^{-4}$ mol/cm^3) with n = 2e$^-$/NaCOOH and assuming $\alpha n_a = 0.5$ and $D = 10^{-5}$ cm^2/s yields a theoretical peak current density of 95 mA/cm^2. The experimental peak current is less than 0.2% of this value, thus formate adsorption rather than diffusion likely limits the reaction rate. It is interesting to note that the reduction of CO_2 past the formate stage to form methanol has been recently reported for the Mo electrode at -0.8 V versus SCE (6, 7). However, the experimental current density (0.12 mA/cm^2) was rather small.

Results obtained with a ruthenium electrode in 1.0 m NaF at pH = 7.7 are presented in Figure 2. The ruthenium metal electrode was smoothly polished, thus the geometrical area (A = 0.40 cm^2) is close to the true electrode area. However, the addition of 0.5 m NaCOOH gave no increase in the cathodic current over that of the background current (dashed line). The complete absence of NaCOOH reduction on this electrode was somewhat surprising since the electrochemical reduction of CO_2 to form methanol and methane has been reported for ruthenium in the form of electroplated and Teflon bonded electrodes (8). The potential of the Pd-H electrode at pH = 7.7 is also indicated in Figure 2, thus, there is an overpotential of nearly 0.3 V before water molecules overcome the activation energy barrier for reduction on Ru in this nearly neutral solution. There is no solid evidence for formate oxidation on the Ru electrode. The solvent oxidation wave near 0.7 V in the presence of NaCOOH in Figure 2 is due to the hydrolysis equilibrium of formate ions that serves as a supplier of the more easily oxidized hydroxide ions. Similar effects are seen with other electrodes.

Various cyclic voltammetric studies of NaCOOH including those shown in Figures 1 and 2 are summarized in Table I. Thirty different electrode materials were investigated in twelve different electrolyte solutions with each electrode being tested in at least two different electrolyte solutions. Twelve electrodes are listed for possible formate reduction, but none of the effects were any larger than that shown for the Mo electrode in Figure 1. Strong formate oxidation was observed on Pt and Pd electrodes, while smaller oxidation effects were found with Rh, Ir, and Au electrodes. The majority of the electrodes tested showed neither formate oxidation nor formate reduction; this suggests very little adsorption of formate on these electrode surfaces.

The electrolytes investigated include metal ions that are reported to form complexes with formate (Mg^{++}, Ca^{++}, and Ba^{++}), anions that show very little specific adsorption (F^-, OH^-, ClO_4^-, NO_3^-, $SO_4^=$), organic anions (CH_3COO^-) and organic cations ($(C_2H_5)_4N^+$). Similar results were often obtained for different electrolytes. For example, the Ag electrode showed no reaction for 0.5 m NaCOOH in 5.0 m LiClO$_4$ + NaOH (pH = 14), 3.0 m La(ClO$_4$)$_3$, 1.0 m Mg(ClO$_4$)$_2$ (T = 80°C), 1.0 m Na$_2$SO$_4$ (T = 60°C) and 0.2 m Na$_2$SO$_4$ (T = 0°C). If an electrode showed evidence for formate reduction in any electrolyte, it was placed in the formate reduction group in Table I. The cobalt electrode showed possible formate reduction in 5.6 m LiClO$_4$ + NaOH (pH = 13) and in 0.1 m (C$_2$H$_5$)$_4$NClO$_4$ (T = 0°C) but tested negative for formate reduction in 3.0 m LaClO$_4$)$_3$, 3.0 m

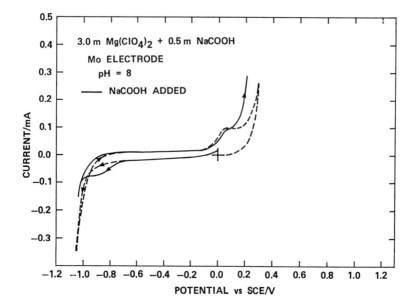

Figure 1   Cyclic voltammogram using a molybdenum electrode
(A = 0.2 cm^2) for 0.5 m NaCOOH added to 3.0 m Mg(ClO$_4$)$_2$ (T =
23°C, ν = 20 mV/s).

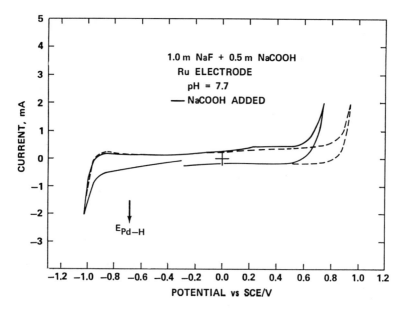

Figure 2.   Cyclic voltammogram using a ruthenium electrode
(A = 0.4 cm^2) for 0.5 m NaCOOH added to 1.0 m NaF (T = 23°C,
ν = 20 mV/s).

$Mg(ClO_4)_2$, 3.9 m $LiClO_4$ (T= 80°C), 1.0 m Na_2SO_4 (T = 60°C) and 12 m KOH (T = 0°C).

Table I.   Summary of Electrochemical Studies of 0.5 m Sodium Formate Using Various Electrodes[a] and Electrolytes.[b,c]

Formate Reduction	Formate Oxidation	No Reaction
Fe, Co, Cu Mo, Rh, In, Sn, Sb W, Ir, Pb, Bi	Rh, Pd Ir, Pt, Au	Glassy C, Al Ti, V, Cr, Ni, Zn Zr, Nb, Ru, Ag, Cd Ta, Re, Hg

[a] Electrodes listed in order of increasing atomic number.
[b] Electrolytes tested: $Mg(ClO_4)_2$, $Ca(ClO_4)_2$, $Ba(ClO_4)_2$, $La(ClO_4)_3$, $LiClO_4$,   NaF, Na_2SO_4, $NaNO_3$, $NaC_2H_3O_2$, $(C_2H_5)_4NClO_4$, Na_2CO_3, KOH.
[b] Most investigations were made at room temperature (23°C). Some experiments were made at 0, 60, and 80°C.

The goal of the various investigations summarized in Table I was to find some electrode-electrolyte-temperature and pH combination that was favorable for the reduction of formate. None of the combinations tested, however, gave formate reduction at useful current densities. In general, formate reduction was more favorable at low temperatures than at high temperatures. This again suggests an adsorption-limited reaction.

Striking effects are found for the platinum electrode when NaCOOH is added to 1.0 m NaF as shown in Figure 3. Large oxidation peaks are observed during both the positive and negative-going potential sweeps due to the oxidation of the formate anion

$$HCOO^- \rightarrow H^+ + CO_2 + 2 \ e^- \tag{4}$$

to form CO_2. The peak current during the negative potential sweep is 5 mA; this is nearly 30% of the theoretical peak current due to diffusion. Even more important is the new cathodic peak near −0.6 V that follows the strong formate oxidation during the negative-going potential sweep. This cathodic peak is likely the reduction of the CO_2 released during the formate oxidation. The observation of this CO_2 reduction peak on a precious metal is made possible by the use of a neutral solution that permits only the reduction of water molecules as a competing reaction. The potential of the Pd-H electrode in 1.0 m NaF at pH = 7.7 is also shown in Figure 3, thus, there is a potential window of over 0.3 V for the Pt electrode between the RHE potential and the actual reduction of water molecules. The reduction of water molecules on platinum is apparently hindered even further by the oxidation of formate and the adsorption of CO_2 on the electrode surface.

Results very similar to those shown in Figure 3 were also observed with 0.50 m NaCOOH in 1.0 m $LiClO_4$, 1.0 m Na_2SO_4 and 1.0 m $NaNO_3$ electrolytes with the platinum electrode. Each of these electrolytes gave a potential window of over 0.3 V between the RHE potential and the actual reduction of water molecules. The reduction of CO_2 following the formate oxidation was observed within this potential window during the negative-going potential sweep. This reduction of CO_2 on platinum was not observed with 0.50 m NaCOOH in 1.0 m electrolyte solutions of $NaHSO_4$, Na_2HPO_4, $NaCO_3$ or KOH. Solvent reduction occurred close to the measured RHE potential for each of these electrolytes. The electrochemical reduction of CO_2 at potentials positive to the onset of the hydrogen evolution reaction is apparently hindered by electrolytes that are too acidic ($NaHSO_4$), too basic (KOH), or act as proton donors or acceptors (Na_2HPO_4, Na_2CO_3). Buffered solutions, therefore, would not give a wide potential window for studies of CO_2 reduction on precious metals such as platinum and ruthenium.

The reduction of CO_2 on high hydrogen overvoltage metals such as Cd, In, Sn, and Pb generally yields only HCOOH or $HCOO^-$ as reaction products (9). The further reduction of formate to methanol is extremely difficult. A better approach for the electrochemical reduction of CO_2 to CH_3OH may be the use of precious metal electrodes. The catalytic properties of such metals are well-known. Results presented in Table I suggests that appreciable formate adsorption occurs only for Rh, Pd, Ir, Pt, and Au electrodes. With good formate adsorption along with the catalytic properties of the precious metal, the reduction of CO_2 may proceed past the formate intermediate on these electrodes. Since these electrodes are also good catalysts for the hydrogen evolution reaction, nearly neutral electrolytes that are poor proton donors or acceptors (unbuffered solutions) such as $LiClO_4$ and Na_2SO_4, should be used. This will force the competing cathodic reaction to be the reduction of water molecules that involves the breakage of H-O bonds rather than the reaction of the more easily reduced acidic or basic species.

The reduction of CO_2 on a Ru-Ir alloy electrode (A = 1.9 cm^2) in 0.2 m $LiClO_4$ at 23°C is shown by the solid line in Figure 4. The bubbling of CO_2 through the electrolyte caused the pH to drop from near neutral to pH = 4.2, thus, the background current (dashed line) was also measured at this pH by introducing a small amount of $HClO_4$ into the solution. In the CO_2-saturated solution, a new reduction wave is observed that begins near -0.5 V. The increase in cathodic current over that of the background current is 0.2 mA (0.1 mA/cm^2) at -0.6 V. The Pd-H potential displayed in Figure 4 shows that a potential window of about 0.3 V is available for CO_2 reduction on the Ru-Ir electrode between the RHE potential and the observed reduction of water molecules in the $LiClO_4$ electrolyte.

Investigations of other precious metal electrodes also showed increased cathodic currents in CO_2-saturated solutions. These studies include Ir, Pd, Pt, and Pt-Ru alloy in 0.2 m $LiClO_4$ and Pd, Pt, and Ru in 1.0 m Na_2SO_4. Although the Ru metal electrode did not give any evidence for a new cathodic peak, there was an increase in cathodic current beginning near -0.7 V for the CO_2-saturated solution. The Ru electrode was inactive with respect to reactions of NaCOOH (Table I), but the reduction of CO_2 to CO, CH_3OH, and even CH_4 has been reported for this metal (6, 8).

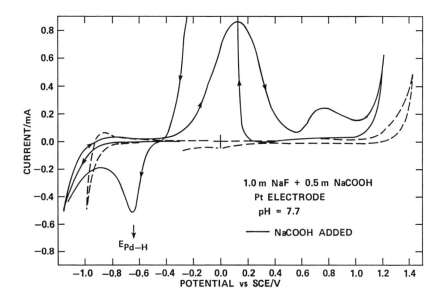

Figure 3   Cyclic voltammogram using a platinum electrode
(A = 0.2 cm^2) for 0.5 m NaCOOH added to 1.0 m NaF (T = 23°C,
ν = 20 mV/s).

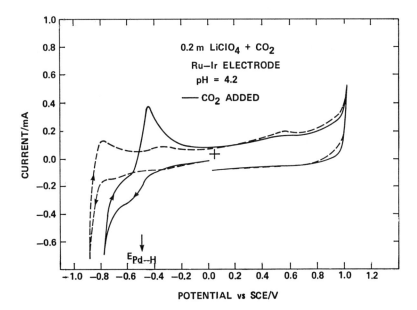

Figure 4.   Cyclic voltammogram using a ruthenium–iridium alloy
electrode (A = 1.9 cm^2) for a CO_2-saturated solution of 0.2 m
LiClO₄ (T = 23°C, ν = 20 mV/s).

Studies are in progress to identify and quantify the products formed by the electrochemical reduction of CO_2 on precious metal electrodes as well as on other electrodes such as Mo when nearly neutral electrolytes are used that minimize proton donor or acceptor reactions. A review of CO_2 reduction on metal electrodes shows that CH_4 is produced on Ru and Cu (8, 9), CH_3OH is a major product on Ru and Mo (6-8), carbon monoxide is formed on Ru, Pd, Pt, Co, Fe, Au, and Ag (7-9), $HCOO^-$ is the main product on Cd, In, Sn, and Pb (3, 9), and a product more complex than formic acid is reported for Pt (10).

Acknowledgments
The authors thank Dan Bliss and George McManis for experimental assistance and helpful discussions. This work was supported by the Office of Naval Research.

Literature Cited

1. Darensbourg, D. J.; Ovalles, C. Chemtech. 1985, 15, 636.
2. Ulman, M.; Aurian-Blajeni B.; Halmann, M. Chemtech. 1984, 14, 235.
3. Russell, P. G.; Kovac, N.; Srinivasan, S.; Steinberg, M. J. Electrochem. Soc. 1977, 124, 1329.
4. Miles, M. H.; Fletcher, A. N.; McManis, G. E. J. Electroanal. Chem., 1985, 190, 157.
5. Gileadi, E.; Kirowa-Eisner, E.; Penciner, J. Interfacial Electrochem., Addison-Wesley: Reading, MA, 1975; pp. 220-224.
6. Summers, D. P.; Frese, Jr., K. W. The Electrochemical Society Extended Abstracts, The Electrochemical Society: Boston, 1986; Vol. 86-1, Abstract 478, p. 697.
7. Summers, D. P.; Leach, S.; Frese, Jr., K. W. J. Electroanal. Chem. 1986, 205, 219.
8. Frese, Jr. K. W.; Leach, S. J. Electrochem. Soc. 1985, 132, 259.
9. Hori, Y.; Kikuchi, K.; Suzuki, S. Chem. Letters, 1985, 1695
10. Vassiliev, Yu. B.; Bagotzky, V. S.; Osterova, N. V.; Mikhailova, A. A. J. Electroanal. Chem. 1985, 189, 311.

RECEIVED February 24, 1987

Chapter 14

Electrochemical Activation of Carbon Dioxide

K. Chandrasekaran and J. O'M. Bockris

Department of Chemistry, Texas A&M University, College Station, TX 77843-3255

The surface states at the semiconductor electrolyte
interface under illumination for the electrochemical
reduction of carbon dioxide has been determined to be
10^{14} cm^{-2}. Surface states are induced by adsorbed
ions and act as faradaic mediators for the photoelectrochemical reduction of carbon dioxide. It is
shown that CO_2 is adsorbed on platinum and adsorbed
CO_2^- is the intermediate radical. The rate determining step involves further reduction of CO_2^- to give
the final products. Adsorption of NH_4^+ ions on p-GaP
has been studied using FTIRRAS. At cathodic potentials adsorbed ammonium ions are reduced and the
reduced ammonium radical desorbs. The structure of
adsorbed ammonium is investigated.

Electrochemical reduction of carbon dioxide provides one method of
converting this plentifully available substance to useful fuels.
It can be carried out biologically (1-2) as in photosynthesis; in
the gas phase (3-4), heterogeneously (5-7); electrochemically (8-15)
or photoelectrochemically (18-20). The efficiencies of the biological and heterogeneous processes are impractically small. Electrochemical reduction of carbon dioxide has been carried out on several
metal electrodes (21-25), although a large overvoltage is required.
Electrocatalysts (26-27) can be used to decrease this overvoltage.
It has been proposed that the slow rate is due to the formation of a
one-electron reduction intermediate, CO_2^-, which is involved in the
rate determining steps (28).

$$CO_2{}_{sol} \overset{\rightarrow}{\leftarrow} CO_2{}_{ads} \tag{1}$$

$$CO_2{}_{ads} + e^- \overset{\rightarrow}{\leftarrow} CO_2^-{}_{ads} \tag{2}$$

$$CO_2^-{}_{ads} + H_2O + e^- \overset{\rightarrow}{\leftarrow} HCOO^- + OH^- \tag{3}$$

The proposed mechanism involves both adsorbed CO_2 and CO_2^- radicals. However, there has been hitherto no direct evidence for the presence of adsorbed CO_2 radical.

Light energy may be used to reduce the necessary electrical potential in photoelectrochemical reactions. The overpotential is decreased by 700 mV for the photoelectrochemical reduction of CO_2 on p-CdTe, compared to that on indium - the best metal electrode for CO_2 reduction. For these semiconductors which involve a high concentration of surface states, the double layer at the semiconductor-electrolyte interface plays an important role in the kinetics of photoelectrochemical reactions. In this paper, we report spectroscopic and impedance aspects of the electrode-electrolyte interface as affected by reactants and radicals involved in CO_2 reduction.

EXPERIMENTAL

Preparation of Electrode

Single crystal CdTe (100) and GaP (110) was cut into 1mm thin wafers. The wafers were wiped clean and sonicated in absolute ethanol for 30 minutes. One side of the wafer was etched with aqua-regia for 30 seconds and washed. An ohmic contact was made with Ga-In alloy. The latter was prepared by mixing equal amounts of Ga and In (wt/wt) at 120°C for 10 minutes under a nitrogen atmosphere to avoid oxide formation. The molten alloy was cooled to room temperature in this atmosphere. Ohmic contact was made at two different positions on the etched face and the resistance between these two contacts was measured. The polarity of the terminals were changed and the resistance was measured again. Etching and rubbing of the Ga-In alloy was continued until the same resistance for current passage in each direction was obtained. After this, an ohmic contact was made on the entire face. The electrode holder was made of copper rod. The rod was covered with teflon to avoid contact with solution. The front side of the electrode was etched for 30 seconds in aqua regia, rinsed with water and immersed into the electrolyte immediately. Before carrying out experiments, the electrode was cycled between -0.56 and -2.24 NHE for c. 20 minutes.

Electrolytes

Tetraalkylammonium perchlorate (TBAP)(Fluka) was recrystallized from ethanol. Dimethylformamide was used without further purification. Triply distilled and pyrolyzed water was used. 0.1 M tetraalkylammonium perchlorate in dimethylformamide - 5% water mixtures were used as electrolyte for the photoelectrochemical reduction of CO_2 to CO.

Cell Compartment

The cell was made up of PYREX glass with an optical quartz window in the front (Fig. 1). Reference and working electrode compartments were fixed on the sides. The working electrode was mounted on the cell by means of a teflon screw for ease of position adjustment. CO_2 was bubbled through chromic acid and DMF to remove impurities, e.g., methanol vapor. A CO_2 blanket was maintained during the

experiments. CdTe electrodes were illuminated with 555 nm light
while GaP electrodes were illuminated with 436 nm light.

Impedance Measurements

Experiments were carried out under potentiostatic conditions using
an 1172 Solartron Frequency Response Analyzer and 1186 Solartron
Electrochemical Interface. A small (input) amplitude (10 mV) sine
wave (P sin ωt) was applied to the system under study. The response
of the system to the applied perturbation was monitored as a sine
wave current [Y sin (ωt + θ_y)] and a sine wave potential [X sin (ωt +
θ_x)]. These were transformed into the complex form A_y + i B_y and
A_x + i B_x, respectively. The real and imaginary parts of the impedance were computed using the relation t = (A_y + i B_y)/(A_x + i B_x)
where the phase shift θ is θ_y - θ_x.

  The AC potential output was measured between the working
electrode and the reference electrode, and the AC current measured
between the working electrode and the counter electrode. Thus, the
impedance between the working electrode and the Luggin capillary was
measured.

  The DC potential of the working electrode was controlled either
by means of the potentiostat or the Frequency Response Analyzer. A
1000 ohm standard resistor was used to measure the DC current. Ten
readings were averaged at each frequency. The frequency range used
was from 0.1 Hz to 9999 Hz. Ten readings were recorded per decade
of frequency with a delay time of 10 sec between readings taken at
each frequency.

FTIR Spectra

A platinum foil was used as a working electrode. It was 10 mm in
diameter, fixed on a polyethylene rod. The tip of the rod was melted
and cooled to provide a leakproof sealing. The electrode was
polished with 0.05μm alumina paste. $LiClO_4$ (0.4M) dissolved in HPLC
grade acetonitrile (Fischer Scientific) was the solvent. The solution was pre-electrolyzed in a nitrogen atmosphere for 2 hours to
remove residual water. The final water content of the solution was
estimated by means of cyclic voltammetry to be 0.01%. Carbon dioxide
was bubbled through the solution for 40 minutes and the cell was
sealed. Similar results were observed when the gas was bubbled
through the solution continuously during the experiment. A platinum
coil was used as the counter electrode and Ag/Ag^+ (a silver wire in
acetonitrile containing 0.1M silver nitrate) was the reference
electrode. The electrode was potentiostatically polarized in the
region of 0.0 to -2.2V NHE (-0.8 to -3.0V vs Ag/Ag^+) and IR spectra
were recorded at eight different potentials in this region.

  A Digilab FTS-20E spectrometer with Nova 4 computer was used to
record the spectra of the adsorbed species using polarization
modulation approach (30). Detailed discussion of the instrumentation
is given elsewhere (30).

. RESULTS

Effect of Surface Treatment

The onset potential for the freshly etched p-CdTe for the photoelectrochemical ·reduction of carbon dioxide is -0.76V NHE. When the
electrode was cycled between -0.56 to -2.24V under illumination, the
onset potential for the photocurrent shifts to less cathodic potentials and remains constant at -0.66V after about c. 20 minutes. When
the electrode is potentiostated at -2.0V the photocurrent remained
constant for about 24 hours. These results are reproducible and consistent with the published data. Surface analysis of the etched
surface of p-CdTe using XPS and SEM showed only trace amounts of carbon and oxygen.
    Several etching procedures were attempted for p-CdTe for the
photoelectrochemical reduction of carbon dioxide. Etching with
dilute thiosulfite or bromine in methanol did not result in better
photocurrent-potential relationship. Hence, it was concluded that
etching with aqua regia followed by rinsing with water is the best
surface treatment for the photoelectrochemical reduction of carbon
dioxide. All the impedance results described below were recorded
using this surface in contact with electrolyte.

Photoelectrochemical Reduction of CO_2 on CdTe

Photocurrent-Potential Relationship. The photocurrent-potential
curve under monochromatic light (555 nm) in a DMF (5% H_2O) solution
containing 0.1M TBAP is shown in Fig. 2. In the absence of CO_2, the
photocurrent starts to increase at -1.4V NHE due to hydrogen evolution. When CO_2 gas is bubbled through the solution, the onset
potential for the photocurrent is shifted to less cathode potentials
by about 700 mV. The reduction product was found to be carbon
monoxide. At low cathodic potentials an anodic current is observed.

Impedance Spectrum of CdTe-DMF (5% H_2O) Containing 0.1 TBAP. When
the real (Z') and imaginary (Z") impedances are recorded under
illumination as a function of frequency from 0.1 to 9999 Hz at -0.76V
NHE, the real part of the impedance is found to decrease with increasing frequency (Fig. 3), while the imaginary part of the impedance passes through a pronounced maximum at intermediate frequencies
and a subsidiary maximum at high frequencies.
    The complex plane plot measured at -1.8V vs NHE is shown in Fig.
4. Parts of the plot can be represented by semicircles. In most
cases, the plots can be divided into three semicircles which, as
shown later, correspond to the dominance of different parts of the
equivalent circuit in different frequency ranges.

Potential Dependence of the Impedance Spectra. In Fig. 5, the
maximum of the Z" - frequency plot is given as a function of potential. The maximum value of Z" decreases as the electrode potential
is made more cathodic.
    Fig. 6 shows the dependence of frequency (f_{max}), at which the
maximum in the Z"-frequency plot occurs, as a function of electrode
potential. As shown, the potential at which f_{max} is observed shifts

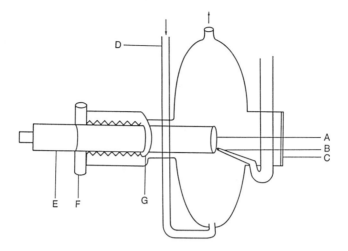

Figure 1.   The electrochemical cell.   A: Semiconductor electrode, B: Luggin capillary, C: Quartz window, D: Gas inlet,
E: Teflon covered copper rod, F: Teflon screw and G: O-ring.

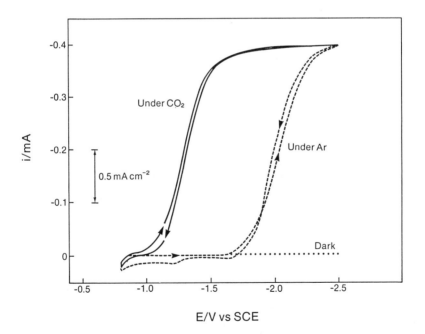

Figure 2.   Photocurrent potential curve for the photoelectrochemical reduction of carbon dioxide on p-CdTe in DMF (5% H_2O)
solution containing 0.1M TBAP.

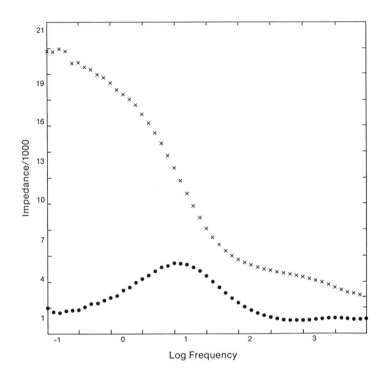

Figure 3.   Bode plot for CdTe - DMF (5% H_2O) interface containing
0.1M TBAP under illuminations (555nm, 2.5mw/cm^2) at -0.76 NHE.
CO_2 1 atmosphere.

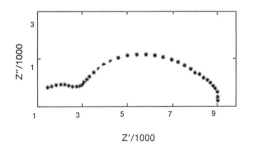

Figure 4.   Complex plane plot for CdTe - DMF (5% H_2O) interface
containing 0.1M TBAP.   Conditions as in Fig. 2.

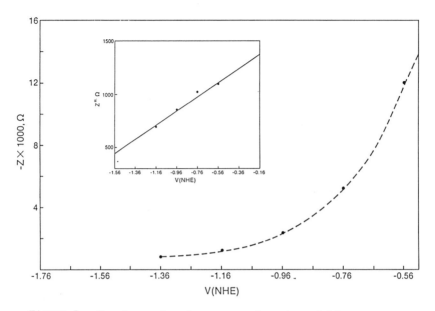

Figure 5.   Imaginary impedance as a function of bias potential for CdTe – DMF (5% H_2O) interface.   Conditions as in Fig. 2.

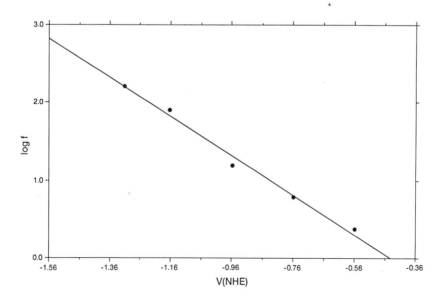

Figure 6.   Potential dependence of imaginary impedance maximum for CdTe electrolyte interface.   Conditions as in Fig. 2.

to higher frequencies with increase in electrode potential in the
cathode direction.

Impedance Spectra in the Presence of Tetraalkylammonium Salts. The
observation of the impedance spectra obtained at the same electrode
potential, -0.96V (NHE), and in the presence of a series of tetraalkylammonium cation shows that the maximum which is observed in the
low frequency region is shifted to lower frequencies as the carbon
chain length is decreased, while the frequency at which the maximum
is observed at higher frequencies remains constant. Z' is observed
to decrease with decrease in the carbon chain length.

Photoelectrochemical Reduction of CO_2 on GaP

Impedance Spectrum of GaP-aqueous DMF Containing 0.1M TEAP. Fig. 7
shows Z" as a function of frequency for p-GaP during the reduction of
CO_2 in aqueous DMF containing 0.1M TEAP. Z" passes through a broad
maximum between 1-100 Hz and a shoulder is observed on the high frequency side of the maximum.
     Z' decreases with an increase in the frequency with an inflec-
tion corresponding to the Z" maximum.

Potential Dependence of Impedance Spectra. Fig. 8 shows the
frequency at which Z" maximum occurs as a function of electrode
potential. Contrary to the behavior in CdTe, where Z" decreases with
increasing cathodic potential, Z" passes through a minimum at -1.16V
(NHE).

Adsorption of CO_2 and CO_2^- on Platinum

The spectra reported here were obtained by subtracting from the
spectra at various potentials, the reference spectrum at 0.0V NHE;
the spectra reported here are thus termed difference spectra. The
difference spectrum of surface adsorbed species on platinum in
acetonitrile at -1.2V NHE is shown in Fig. 9. The peaks pointing
downward show a decrease in surface concentration with respect to the
reference potential, 0.0V NHE, and the peaks pointing upwards indicate an increase in surface concentration. Three peaks pointing
downwards and one peak pointing upwards in the region 2400-1500 cm^{-1}
are seen in Fig. 9.
     There is a broad maximum centered around 1680 cm^{-1}, the
intensity of which increases in the cathodic direction. Absorbance
at 1680 cm^{-1} as a function of electrode potential is shown in Fig.
10. Full width at half maximum for this peak is 40 cm^{-1}. The
relative area under the peak has been taken as proportional to the
concentration of adsorbed species on the surface of the electrode, an
assumption which is acceptable at least up to $\theta = 0.5$. The area of
the peak in the region 1650-1700 cm^{-1} is shown in Fig. 11 as a function of electrode potential. Adsorption increases in the cathodic
direction and reaches a saturation value around -2.0V NHE.
     A sharp peak centered at 2342 cm^{-1} is also observed and may be
identified (31) with adsorbed CO_2 (Fig. 12).
     Absorbance corresponding to the $C\equiv N$ stretching vibrations of
adsorbed acetonitrile (32) is observed at 2250 cm^{-1}. The spectra of
adsorbed CO_2 and adsorbed CH_3CN at various electrode potentials are

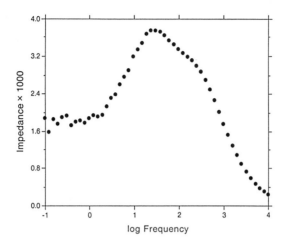

Figure 7.   Impedance spectrum for GaP – DMF (5% H_2O) containing 0.1M TEAP under illumination (436 nm) at –0.76V NHE. CO_2 1 atmosphere.

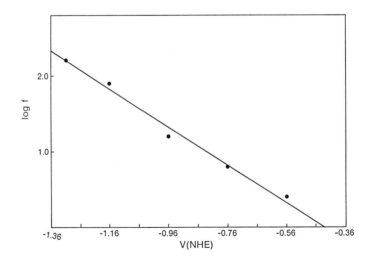

Figure 8.   Frequency of imaginary impedance maximum as a function of bias potential of GaP electrolyte interface containing CO_2 under illumination.

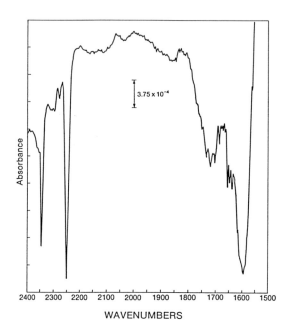

Figure 9. The differential spectrum of surface adsorbed species on platinum in acetonitrile at -1.2V NHE.

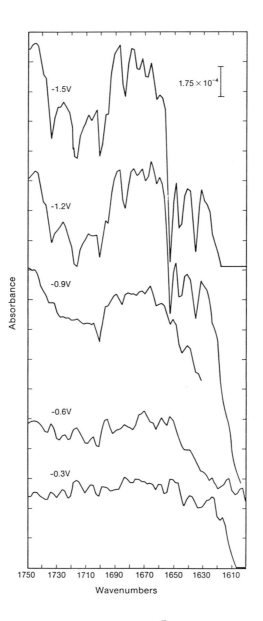

Figure 10.   Absorption spectrum of CO_2^- radical adsorbed on platinum in acetonitrile containing 0.4M $LiClO_4$.

shown in Fig. 12. The area of the peak centered at 2250 cm^{-1}
decreases with an increasingly cathodic potential.

Adsorption of NH_4^+ on GaP

Ammonium ions in acetonitrile solution show a broad maximum around
3400 cm^{-1} corresponding to the symmetric N-H stretching and a sharp
band around 1600 cm^{-1} for N-H asymmetric stretching (33). Ammonium
ions adsorbed on GaP shows a broad maximum centered at 3475 cm^{-1} and
3000 cm^{-1} (Fig. 13). There are two maxima corresponding to N-H
deformation in the region 1700 cm^{-1} and 1560 cm^{-1} (Fig. 13). The
potential dependence of all these four peaks follow the same trend
(Fig. 14). Absorbance increases in the beginning and then decreases
at more cathodic potentials.

DISCUSSION

The Equivalent Circuit

An appropriate equivalent circuit can be created if the sequence of
events between the creation of hole-electron pairs and (e.g.), the
acceptance of electrons in solution is clearly known. The principal
arbiter of an equivalent circuit is the degree to which it represents
trends in impedance as a function of frequency.
    An equivalent circuit for a photoelectrochemical system has to
take into account: 1. The generation of electron-hole pairs; 2.
Passage of carriers through the space charge region; 3. Passage of
carriers through the surface states; 4. Passage of carriers through
the double layer; 5. The circuit must allow for the fact that both
holes and electrons are generated but move in opposite directions.
    Considering the sequence of events at the semiconductor-
solution interface, the four circuits shown in Fig. 15 were all used
to simulate the results. It is seen that the circuit 15d fits the
results to a greater degree than do other circuits. It is reasonable, therefore, to conclude that the appropriate circuit for the
evaluation of N_{ss} (the surface state concentration per square cm) is
15d.

Evaluation of Parameters

The total impedance of the circuit given in Fig. 15 for the circuit
15d is given by the following equations.

$$Z' = \frac{R_{sc}}{1+\omega^2 C_{sc}^2 R_{sc}^2} + \frac{R_{ss}}{1+\omega^2 C_{ss}^2 R_{ss}^2} + \frac{R_{DL}}{1+\omega^2 C_{DL}^2 R_{DL}^2} + R + R_{so}, \qquad (4)$$

$$Z'' = \frac{R_{sc}^2 C_{sc}}{1+\omega^2 C_{sc}^2 R_{sc}^2} + \frac{R_{ss}^2 C_{ss}}{1+\omega^2 C_{ss}^2 R_{ss}^2} - \frac{R_{DL}^2 C_{DL}}{1+\omega^2 C_{DL}^2 R_{DL}^2} \qquad (5)$$

where C_{sc} is the space charge capacitance, C_{ss} is the surface state
capacitance, C_{DL} is the double layer capacitance, R_{sc} is the space
charge resistance, R_{ss} is the surface state resistance, R_{DL} is the

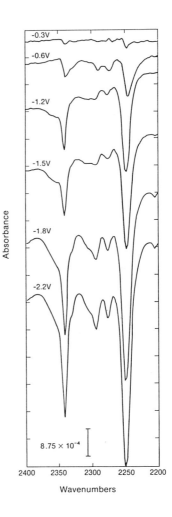

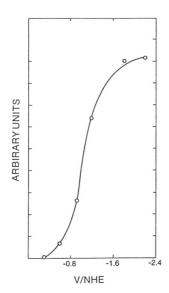

Figure 11. Relative concentration of adsorbed CO_2^- on platinum (see text).

Figure 12. Adsorption spectra of adsorbed CO_2 (2340 cm^{-1}) and CH_3CN (2250 cm^{-1}) on platinum in acetonitrile containing 0.4M $LiClO_4$.

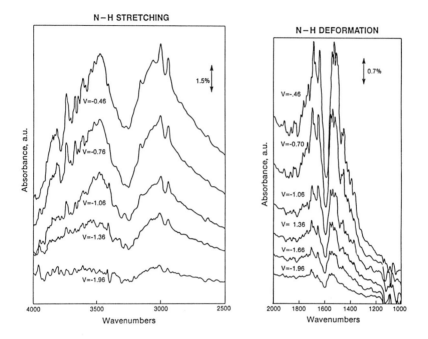

Figure 13. Absorption spectrum of ammonium ion (N-H stretching and N-H deformation) on GaP in acetonitrile.

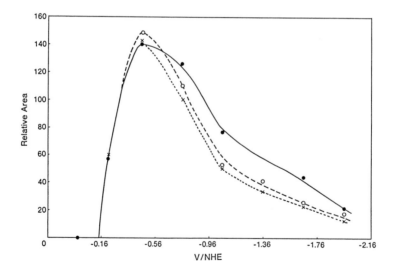

Figure 14. Potential dependence of ammonium ion adsorption on GaP. (.) 3475 cm^{-1}, (o) 1700 cm^{-1} and (x) 1560 cm^{-1}.

double layer resistance, R_{so} is the solution resistance and ω is the frequency. The Z"-frequency plot should pass through three maxima. In fact, in a majority of the experiments, only two maxima were observed, although in a minority, three maxima occurred.

Since the Z" maxima in the Bode plots are well separated, the impedance contribution of one to another can be taken as negligible. At $Z"_{max}$, $\omega CR = 1$, hence $Z"_{max} = R/2$ and $C = 1/\omega R$. Capacitances and resistances calculaed from complex plane plots (Fig. 4) give numerical values of C&R which differ by less than 20% with those from the Z&Z'-frequency relations.

Capacitances and resistances calculated for the two maxima at various bias potentials are shown in Table 1. Capacitances calculated from the high frequency maximum are of the order of 0.1 μF cm^{-2} and increase with increasing cathodic potential. The time constant remains constant in the potential region studied. The value of the capacitance and the potential dependences are close to those expected for the space charge region and hence are assigned to this.

Table I . Capacitances and Resistances for CdTe in DMF (5% H_2O). Containing 0.1 M TBAP

V (NHE)	R_{ss} ohms	C_{ss} μF	R_{sc} ohms	C_{sc} μF	R_{DL} ohms	C_{DL} μF
-0.56	2940	421.6	240	0.331	500	100
-0.76	2880	421.6	300	0.265	250	100
-0.96	1440	341.6	288	0.276	150	100
-1.16	360	133.3	204	0.390	63	100
-1.36	156	30.0	132	0.693	28	100
-1.56	84	18.33	85	0.948	23	100

The capacitances corresponding to the low frequency maximum are of the order of 10-400 μF cm^{-2} and decrease in magnitude with increasingly cathodic potential. Resistances corresponding to these maxima are of the order of 1000 ohms and decrease with increasing cathodic potential. Since most of the capacitance values are higher than those characteristic of the double layer, and vary with potential, they may be attributed to surface states.

The surface state resistance is larger than the space charge resistance in the Tafel region. Were the rate-determining step for the photoelectrons lie in the space charge region and not at the interface, the resistance of the space charge region would be greater than the value for the surface states. Hence, the rate determining step lies not in the space charge region, but at the interface. At sufficiently cathodic potentials, the space charge resistance does become relatively greater than other series resistances, consistent with the conclusion that this region becomes rate determining at high current densities (i.e., at sufficiently high current densities, a transport controlled limiting current is observed (34).

Calculation of Sruface State Capacitance at a Fixed Potential

The number of surface states at a given electrode potential can be
calculated without model assumptions from the differential surface
capacitance, using the relation (35).

$$e_o N_{ss} = Q_{ss} = \int_{V_0}^{V_1} C_{ss} dV$$

where N_{ss} is the number of surface states per unit area at the
potential V_1, C_{ss} is the differential surface state capacitance and
V_0 is the potential at which the surface state capacitance is zero
(in practice, the minimum value on the C_{ss} potential plot).
    The surface state capacitance for the CdTe-electrolyte inter-
face is plotted as a function of electrode potential in Fig. 16
(the minimum was taken as the value at 0.2V NHE). The surface state
capacitance decreases in the cathodic direction in the region -0.56
to -2.26V (NHE). Capacitance measurements at cathodic potentials
less negative than -0.56V could not be carried out because of the
onset of a CO_2-independent anodic dark current. Assuming (in con-
sistence with other examples of pseudo capacitance behavior) that
the capacitance-potential curve is symmetrical with respect to a
maximum at -0.66V, the number of surface states was calculaed using
the above equation. The number of surface states as a function of
electrode potential, on the basis of this assumption, is shown in
Fig. 17. Geometric area of the electrode was used to calculate the
surface state density. Real surface area may be larger.
    As to the nature of the surface states represented by Fig. 17,
these are not likely to be calssical surface states provided, e.g.,
by dangling bonds. Such bonds (characteristic of the semiconductor-
vacuum interface) are likely to have been removed by the adsorption
of water from the solution. They would not have been expected to
vary with potential (cf. Fig. 17). The surface states being
measured here may result from adsorption of ions from solution.
These kinds of surface states are expected to vary with potential,
solvent, electrode, electrolyte and current density. Adsorption of
the tetraalkylammonium ion in acetonitrile on p-silicon has been
studied by FTIR relection-absorption spectroscopy (30). The adsorption isotherm resulting from such measurements is similar in
nature to the surface state density data for this ion on CdTe. The
fact that the measured surface state density on CdTe varies with the
nature of the cation (Table 2) is consistent with the concept that
the states arise from ionic adsorption.

Table II.   Surface State Capacitance for Several Electrolytes at the
             CdTe-DMF (5% H_2O) Interface

Electrolyte	$C_{ss} \mu F \ cm^{-2}$
Et_4NClO_4	32
Pr_4NClO_r	30.4
Br_4NClO_r	16.0
Ot_4NClO_4	4.6

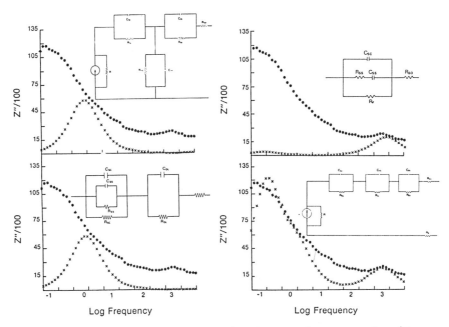

Figure 15. Simulated imaginary impedance – frequency plot (C_{sc} = 0.276 F, R_{sc} = 0.288 k , C_{ss} = 341 F, R_{ss} = 1.44k , C_{DL} = 100 F, R_{DL} = 250) (X) and experimental results (*).

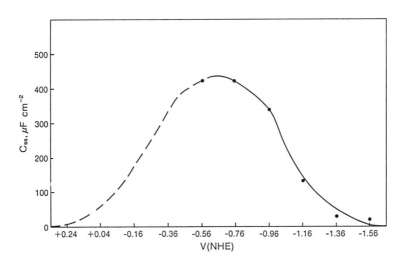

Figure 16. Surface state capacitance as a function of bias potential for the CdTe electrolyte interface.

Thus, in this model, the surface states formed by the tetra-alkylammonium cation would be acting as faradaic mediators. Photo-generated electron passes through the space charge region to surface states, and then to acceptor species in solution. Thus, for the region of potential below the transport controlled limiting current (Table 1), e.g., at -0.96V NHE, the surface state resistivity (cm^{-2}) is some five times greater than the correspondingly resistance in the space charge region. A possible model consistent with these facts would have adsorption of tetraalkylammonium ion (i.e., the formation of surface states) as the rate determining step. The subsequent electron transfer to CO_2 is evidently rapid in this case. Taniguchi et al. (36) have found that addition of ammonium ion to the solution caused significant catalysis of the photoelectrochemical reduction of carbon dioxide and suggested the mechanism.

$$NH_4^+ + e^- \rightarrow NH_4^{\cdot}$$

$$NH_4^{\cdot} + CO_2 \rightarrow NH_4^+ + CO_2^-$$

Such a mechanism would be consistent with the present results.

The absence of the third maximum can be shown to be consistent with the present picture. Thus, if the exchange current density is 10^{-5} A cm^{-2}, $C_{DL} = 60.10^{-6}$ cm^{-2}, then at w = 0.1 HZ, Z" = 20 ohms cm^{-2}, assuming $R_{DL} = RT/i_o F$, the equilibrium value. The value would in fact be less because at -0.96V (NHE), the electrode is 600 mV negative to equilibrium and R_{DL} would become negligible compared with the surface state resistance.

It follows that the reciprocal of R_{ss} would be proportional to the measured photocurrent. Such a relation is shown in Fig. 18 and is consistent with the model suggested.

Effect of Electrolytes

Perchlorate salts of tetraalkylammonium ions were chosen as electrolytes for this study, because they reduce hydrogen evolution. The chain length of the alkyl group was varied, i.e., ethyl, propyl, butyl, and octyl. A decrease in photocurrent was observed when the carbon chain length was increased. The surface state resistance was found to increase with increase of chain length. Correspondingly, a decrease in surface state capacitance was observed. These results indicate that tetraethylammonium ions are adsorbed stronger than tetrabutylammonium ions.

The anamolous behavior of tetraalkylammonium ions can be explained as follows. Hydration of these organic ions may be weak. In DMF (5% H_2O) solution, the solubility of tetraalkylammonium ion increases with increase of alkyl chain length and hence, the ad-sorption of cation decreases with increase of chain length.

Effect of Solvent

When the water concentration in the DMF water mixture is increased from 1%-25%, the surface state resistance decreases and there is a concurrent increase of surface state capacitance. Such a change would be expected to occur if there were adsorption of ions when the water concentration is increased, and this is indeed the case (37).

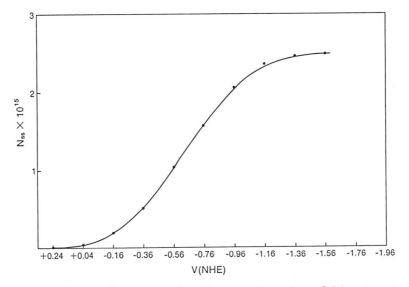

Figure 17. Surface state density as a function of bias potential calculated from the capacitance data.

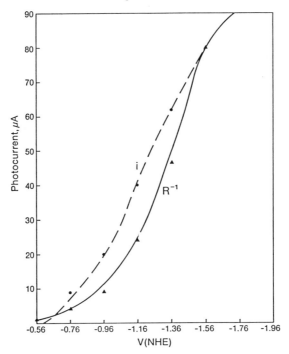

Figure 18. Relative rate constant for the photoelectron transfer across the surface state as a function of bias potential. Photocurrent measured under identical conditions is shown in dotted lines.

GaP-electrolyte Interface

The impedance data for the GaP-electrolyte interface can be represented by the equivalent circuit discussed for the CdTe electrode.
The surface state capacitance calculated by a similar procedure is
shown in Fig. 19 as a function of bias potential. Fig. 20 shows the
surface states density as a function of bias potential. Surface
states density is an order of magnitude less than that at the CdTe
interface under similar conditions. This is consistent with the fact
that the photocurrent for GaP is less compared to CdTe for the[36]
photoelectrochemical reduction of CO_2 (cf. the model suggested of
mediator surface states). The surface state density increases from
its minimum at 0.2V NHE, a potential which lies close to the pzc
value determined for the system (0.17V NHE). Correspondingly, the
form of the surface state density as a function of bias potential
resembles an adsorption isotherm. These results support the concept
that surface dates are induced by adsorbed ions at the interface.

Adsorption of the Carbon Dioxide Radical

The absorption maximum at 1680 cm^{-1} can be attributed to the
adsorbed CO_2 radical. The IR spectrum of this radical has been
recorded at -190°C. It has a sharp maximum at 1671 cm^{-1} (38). The
full width at half maximum is 3 cm^{-1}. The CO_2 radical in the latter
case was generated by the radiolysis of sodium formate in potassium
bromide matrix. The broadening of the spectrum (full width at half
maximum 40 cm^{-1}) is consistent with the model of a radical adsorbed
on the electrode surface.

The integrated peak areas between 1650-1700 cm^{-1}, at several
bias potentials, are shown in Fig. 11. The adsorption of the anion,
CO_2^-, increases in the cathodic direction. If the adsorbed CO_2^-
radial were in equilibrium with CO_2^- in solution a decrease in
adsorption coverage would be expected when the potential is moved in
the cathodic direction. However, increase in anion concentration at
cathodic potentials is consistent with CO_2^- as an intermediate radical
in the electrochemical reduction of CO_2. Thus, from (1)-(3),

$$\theta_{CO_2} = kP_{CO_2}$$

and with (2) in equilibrium

$$\theta_{CO_2^-} = k_2 k_1 P_{CO_2} \, e^{-VF/RT}$$

where V is the electrode potential, k_1 and k_2 are equilibrium
constants for (1) and (2), respectively. Thus, $\theta_{CO_2^-}$ increases with
increasing negative value of V (Fig. 10), with (3) rate determining,
the current density, i,

$$i = 2Fk_3 \theta_{CO_2} - e^{-VF/RT}$$

$$= 2Fk_1 k_2 k_3 P_{CO_2} \, e^{-(1+\alpha)VF/RT}$$

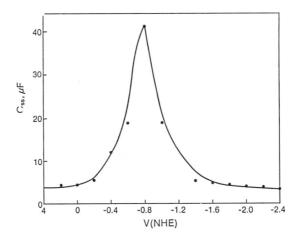

Figure 19. Surface state capacitance as a function of applied bias potential for GaP-DMF (5% H_2O) interface for the photo-electrochemical reduction of CO_2.

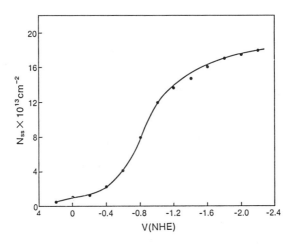

Figure 20. Surface state density as a function of bias potential for GaP-electrolyte interface.

Thus, it is possible to obtain information on the rate-controlling step by spectroscopically identifying an intermediate radical and following the relation of the surface coverage with potential.

Adsorption of Carbon Dioxide

The IR spectrum of CO_2 in gas phase has a maximum of 2349 cm^{-1} (30). Carbon dioxide adsorbed on platinum in acetonitrile has a maximum at 2342 cm^{-1}. Since the frequency shifts (compared to gas phase CO_2) are very small, CO_2 is probably physisorbed. It has been reported that adsorption of CO_2 on platinum at anodic potentials involves chemisorbed CO (39). However, such a chemisorbed CO was not observed at cathodic potentials on platinum in acetonitrile.

If CO_2 is adsorbed parallel to the surface of the electrode, the asymmetric stretching vibrations are not IR active, for the electric field vector of the parallely polarized light is zero in the plane parallel to the surface of the electrode. Since adsorbed CO_2 absorbs at 2340 cm^{-1}, the adsorption occurs through one oxygen atom and the other oxygen atom must project towards the solution (Fig. 21). Symmetric stretching vibrations of CO_2 are not IR active as the transition dipole moment for this symmetric stretching is zero. The bending vibrations of CO_2 occurs at 667 cm^{-1}, which is beyond the sensitivity of the instrument.

The relative concentration of CO_2 decreases in the cathodic direction (Fig. 11). The decreasing concentration of CO_2 may be due to the reduction of CO_2 to give CO_2^-, which absorbs at 1680 cm^{-1}. Adsorption of neutral molecules as a function of electrode potential generally passes through a maximum near the potential of zero charge. The potential of zero charge for platinum acetonitrile interface has been determined to be 0.3V NHE (40). The concentration of adsorbed CO_2 will decrease on either side of potential of zero charge, which is consistent with the present experimental results.

Adsorption of Acetonitrile

The absorption maximum at 2250 cm^{-1} is attributed to the stretching vibrations of C≡N group of acetonitrile (31). Since -C≡N groups parallel to the surface of the electrode are not IR active and C-C-N is linear, the acetonitrile must be adsorbed through the N atom and the methyl group projects towards the solution. Another possibility is that the -C≡N group adsorbs parallel to the surface of the electrode and the stretching vibrations of -C≡N group observed is due to electrochemical Stark effect. But, according to Pons et al. (41), the electrochemical Stark effect is important only at high potentials (10^8 V cm^{-1}) whereas the electric field under the present experimental conditions is less than 4 x 10^7 V cm^{-1}. The electric field increases toward more cathodic potentials. If the electrochemical Stark effect were responsible for the -C≡N vibrations, the intensity would be expected to increase towards more cathodic potentials as the electric field increases, wherreas a decrease in absorbance is observed under these conditions. Hence, the electrochemical Stark effect is unlikely under the present experimental conditions.

The relative concentration of CH_3CN on the surface of platinum decreases with increasing cathodic potentials, because adsorption of neutral molecule decreases away from the potential of zero charge. The acetonitrile concentration will be further decreased at high cathodic potentials by increasing $\theta_{CO_2^-}$. Since CO_2 and CH_3CN are linear, the area occupied per molecule on the surface of the electrode is small. However, once CO_2 is reduced to give CO_2^-, the radical is no longer linear (42)(the 0-C-0 angle is 134°), and the bond length for C-0 is increased from 1.8 Å for CO_2 to 2.1 Å for CO_2^- (42).

The CO_2^- radical may be adsorbed through one oxygen atom or through two oxygen atoms. In either case, the area occupied per CO_2^- ion is larger than that of CO_2. Hence, the adsorbed acetonitrile molecule is expelled increasingly from the surface during process (2) to accommodate the larger CO_2^- radical.

Adsorption of Ammonium Ions

Ammonium ions have been shown to act as a catalyst for the photo-electrochemical reduction of carbon dioxide to carbon monoxide (36). It has been proposed that they are adsorbed on the surface and act as electron mediators for the photoelectrochemcial reduction of CO_2.

Ammonium ions in solution, being symmetrical, shows a broad maximum around 3400 cm^{-1}, corresponding to N-H symmetric stretching and a sharp maximum around 1600 cm^{-1} due to N-H deformation vibrations (32). Ammonium ions adsorbed on GaP show two broad peaks, centered around 3475 cm^{-1}, and 3000 cm^{-1}, corresponding to the N-H symmetric stretching (Fig. 13). Two peaks of equal intensity were observed for N-H deformation vibration: one at 1700 cm^{-1} and another at 1560 cm^{-1} (Fig. 12). Two different N-H groups are involved. From the equal intensities of the two peaks, it is evident that each peak corresponds to two different N-H deformation vibrations.

Three orientations of adsorbed ammonium ions are possible as shown in Fig. 22. Unlike ammonium ions in homogeneous solution, the adsorbed ammonium ion is not a symmetrical tetrahedron. Two different N-H groups are present for the adsorbed ammonium ion. The peaks at 3000 cm^{-1} and 1560 cm^{-1} are attributed to stretching and deformation vibrations of N-H group projecting towards the solution as it lies close to the ammonium ions in solution. Absorbance of 3400 cm^{-1} and 1700 cm^{-1} may be due to adsorbed N-H group. Since the intensities of these two peaks are equal at all bias potentials, it is proposed that the ammonium ion adsorption occurs through two hydrogen atoms as shown in Fig. 21b. For the other two orientations (Fig. 22), the intensity ratios should be 3:1 and 1:3. Symmetric stretching vibrations of the adsorbed ammonium ions also indicate a similar behavior.

The potential dependence of the ammonium ion is anamolous in that it begins at 0.3V negative to the PZC, passes through a maximum, and then decreases with increasing cathodic potential. This behavior may reflect the large dipole moment (3.92 D) of the acetonitrile molecule. It may be necessary to invoke a possible chemical bonding of CH_3CN to GaP. Adsorbed acetonitrile may be reduced at about -0.3V NHE, decreasing the surface concentration, and allowing ammonium ion adsorption. However, at sufficiently cathodic potentials

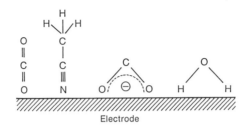

Electrode

Figure 21. Possible structures of adsorbed molecules on platinum.

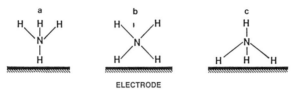

ELECTRODE

Figure 22. Possible structures of adsorbed NH_4^+ ions on GaP.

> -0.7V, ammonium ion is reduced (36), and the reduced radical
decomposes.

$$NH_4^+ + e^- \rightarrow NH_4 \rightarrow NH_3 + H$$

Thus, the ammonium ion concentration decreases at more negative
potentials. Since N-H vibrations of NH_3 are not observed under
our experimental conditions, it is proposed that the decomposition
reaction of ammonium ion occurs in the diffuse layer.

Conclusions

1. Surfce states at the semiconductor-electrolyte interface under
    illumination can be calculated from the impedance measurements
    using the new equivalent circuit proposed.
2. Surface states density at a given bias potential can be calcu-
    lated from integral surface state capacitance.
3. Surface states are induced by adsorption of ions at the
    semiconductor-electrolyte interface.
4. Surface states act as faradaic mediators for the photoelectro-
    chemical reduction of CO_2.
5. Adsorbed CO_2 amd CO_2^- are involved in the electrochemical
    reduction of CO_2.
6. The rate determining step has been determined to be further
    reduction of CO_2^- to give products.
7. Ammonium ions are adsorbed at the semiconductor electrolyte
    interface and the reduced ammonium ion radical acts as mediator
    for the photoelectrochemical reduction of CO_2.

Acknowledgments

The authors would like to thank the Gas Research Institute for supporting this project and Dr. Kevin Krist, Dr. M. A. Habib and Dr. B. Scharifker for discussions.

Literature Cited

1. Bolton, J. R. - Science 202, 705 (1978).
2. Bolton, J. R. and Hall, D. O. - Ann. Rev. Energy 4, 353 (1979).
3. Eremin, E. N., Maltsev, A. N. and Ivanter, V. C. - Zh. Fiz. Khim. 54, 150 (1980).
4. Eremin, E. N., Maltsev, A. N., Ivanter, V. C. and Belova, V. M. - Zh. Fiz. Khim. 55, 1365 (1981).
5. Inoue, T., Fujishima, A., Konishi, S., and Honda, K. - Nature 277, 637 (1979).
6. Halmann, M. - "Energy Resources through Photochemistry and Catalysis" ed. M. Gratzel, Academic Press, New York (1983).
7. Chandrasekaran, K., and Thomas, J. K. - Chem. Letts. 99, 7 (1983).
8. Teeter, T. E., and V. Rysselberghe - J. Chem. Phys. 22, 759 (1984).
9. Russel, P. G., Kovac, N., Srinivasan, S., and Steinberg, M. - J. Electrochem. Soc. 124, 1329 (1977).
10. Udupa, K. S., Subramanian, G. S., and Udupa, H. V. K. - Electrochim. Acta 16, 1593 (1971).
11. Kaiser, V., and Heitz, E. - Ber. Bunsenses 77, 818 (1973).
12. Takahashi, K., Kiralzuka, K., Sasaki, H. and Toshima, S. Chem. Lett. 305 (1979).
13. Meshitsuka, S., Ichikawa, M., Tamura, K. -J. Chem. Soc., Comm., 158 (1974).
14. Fisher, J., Lehmann, T., Hietz, E. - J. Appl. Electrochem. 11, 743 (1981).
15. Geambino, A., and Silvestri, G. - Tetrahedran Lett. 3025 (1973).
16. Halmann, M. - Nature, 275, 29 (1979).
17. Hemminger, J. C., Carr, R. and Somorjai, G. A. - Chem. Phys. Lett. 57, 100 (1978).
18. Taniguchi, I., Blajeni, B. A. and Bockris, J. O'M. - J. Electroanal. Chem. 157, 179 (1983).
19. Taniguchi, I., Blajeni, B. A. and Bockris, J. O'M. - Electrochimica Acta 29, 923 (1984).
20. Bradley, M. G., Taysak, T., Graves, D. G. and Vlalchopoulos, N. A. - J. Chem. Soc., Chem. Comm. 249 (1983).
21. Frese, K. W., Jr. and Summens, D. P. - GRI Annual Report (1985).
22. Liever, C. M. and Lewis, N. S. - J. Am. Chem. Soc. 106, 5083 (1984).
23. Sears, W. M. and Morrison, S. R. - J. Phys. Chem. 89, 3295 (1985).
24. Vassiliev, Y. B., Bagotzky, V. S., Osetrova, N. V., Khazova, O. A. and Magorova, N. A. - J. Electroanal. Chem. 189, 271 (1985).
25. Vassiliev, Y. B., Bagotzky, V. S., Khazova, O. A. and Mayorova, N. A. - J. Electroanal. Chem. 189, 295 (1985).

26. Kapusta, S. and Heckerman, N. - J. Electrochem. Soc. 130, 607 (1983).
27. Kapusta, S. and Heckerman, N. - J. Electrochem. Soc. 131, 1511 (1984).
28. Paik, W., Anderson, T. N. and Eyring, H. - J. Phys. Chem. 46, 3278 (1972).
29. Bockris, J. O'M. and S. U. M. Khan, J. Electrochem. Soc. 132, 2648 (1985).
30. Chandrasekaran, K. and Bockris, J. O'M. - Surface Science 175, 623 (1986).
31. Durana, J. F. and Manty, A. W. - "Fourier Transform Infrared Spectroscopy: Applications to Chemical Systems," Vol. 2; eds. J. R. Ferrano and L. J. Basile, Academic Press, New York (1979).
32. Craver, C. D. - ed. "Desk Book of Infrared Spectra," 2nd ed., The Coblenty Society, Inc., Kirkwood (1980).
33. Socrates, G. - "Infrared Characteristic Group Frequencies," John Wiley & Sons, New York (1980).
34. Sklarczyk, M. and Bockris, J. O'M. - J. Phys. Chem. 9, 831 (1984).
35. Grahame, D. C. - Chem. Rev. 41, 441 (1974).
36. Taniguchi, I., Blajeni, B. A. and Bockris, J. O'M. - J. Electroanal. Chem. 161, 385 (1983).
37. Damaskin, B. B., Petrii, O. A. and Batrakov, V. V. - "Adsorption of Organic Compounds on Electrodes" Plenum Press (1971).
38. Harman, K. O. and Hisatsune, I. C. - J. Chem. Phys. 44, 1913 (1966).
39. Beden, B., Bewick, A., Ragaq, M. and Weber., J. - J. Electroanal. Chem. 139, 202 (1982).
40. Petrii, O. A. and Khomchenko - J. Electroanal. Chem. 106, 277 (1980).
41. Korgeniewski, C., Shirts, R. B. and Pons, S. - J. Phys. Chem. 89, 2297 (1985).
42. Pacanski, J., Wahlgren, H. and Bagus, P. S. - J. Chem. Phys. 62, 2740 (1975).

RECEIVED February 13, 1987

INDEXES

Author Index

Allen, Leland C., 91
Ayers, W. M., 147
Barer, Sol J., 1
Bauch, Christopher G., 26
Bockris, J. O'M., 179
Bolinger, C. Mark, 52
Bruce, Mitchell R. M., 52
Butler, James N., 8
Chandrasekaran, K., 179
Dai, C. H., 133
Darensbourg, Donald J., 26
DuBois, Daniel L., 42
Dumesic, J. A., 102
Ekerdt, John G., 123
Farley, M., 147
Fletcher, A. N., 171

Frese, Karl W., Jr., 155
Freund, H.-J., 16
Jackson, Nancy B., 123
Megehee, Elise, 52
Messmer, R. P., 16
Meyer, Thomas J., 52
Miedaner, Alex, 42
Miles, M. H., 171
O'Toole, Terrence R., 52
Ovalles, Cesar, 26
Rethwisch, D. G., 102
Silver, Ronald G., 123
Stern, Kenneth M., 1
Sullivan, B. Patrick, 52
Summers, David P., 155
Thorp, Holden, 52
Worley, S. D., 133

Affiliation Index

Auburn University, 133
Chem Systems Inc., 1
Electron Transfer Technologies Inc., 147
General Electric Company, 16
Harvard University, 8
Naval Weapons Center, 171
Princeton University, 91

Solar Energy Research Institute, 42
SRI International, 155
Texas A&M University, 26,179
University of Iowa, 102
University of North Carolina, 52
University of Texas, 123
University of Wisconsin, 102

Subject Index

A

Acetaldehyde, 139
Acetone, 139
Acetonitrile
  adsorption, 200
  relative concentration, 201
Acid–base reactions, 11
Acidity constant, 12

Acrylic acid formation, 61
Activation catalysis, 28
Activation energies, 112
Activation energy barrier
  electronic rearrangements, 99
  Michaelis complex formation, 95
  reduction of ruthenium, 173
Activation of CO_2, 131f
Activation parameters, 31

Adsorption
  CO_2 and CO_2^- on platinum, 186
  NH_4^+ on GaP, 190
  nitric oxide, 103
Alkali metals, 134
Alkyl formate production, 33
Aluminum oxide, 102–122
Amino acid side chains, 101
Ammonia plants, 3
Ammonium ion
  absorption spectrum, 192*f*
  acetonitrile solution, 190
  adsorption, 201
  catalyst, 201
  decomposition, 202
  orientations, 201
  potential dependence, 192*f*
Anionic alkyl complexes, 31
Anionic carbonyl complexes, 35
Anionic halide complexes, 37
Aprotic solvents, 13
Aqueous electrolytes, 12
Around the corner S_N2 reaction, 96,99
Associative mechanisms, 83
Auger electron spectroscopy, 164

B

Band gap irradiation, 57
Bent bonds, 17
Beverages, 2
Bicarbonate complex, 74
Bicarbonate decomposition, 131
Bicarbonate ions, 161
Bicarbonate species, 125
Bidentate carbonate I, 130
Bimolecular catalyzed reactions, 92
Bimolecular one-electron steps, 83
Bipolar membrane electrode
  configuration, 148
Bond pairs, 17
Bonding interaction, 21
Bonding modes, 58
Breakage of H–O bonds, 176
Buffered solutions, 176
Bulk electrolysis, 77,78

C

C_1 hydrocarbons, 130
C_2O formation, 96
Capacitance behavior, 194
Capacitances and resistances for
  CdTe, 193*t*
Carbon dioxide
  adsorption
    platinum in acetonitrile, 200
    vibrations, 200

Carbon dioxide–*Continued*
  adsorption and reaction, 123–132
  adsorptive and catalytic
    properties, 102–122
  and bicarbonate, 151
  anion, 19
  aqueous solutions, 171–178
  commercial uses, 27
  economics, 1–7
  electrochemical activation, 179–204
  electrochemical reduction, 155–170,179
  electrostatic repulsion, 147
  equilibria, 8–15
  highly stable molecule, 26
  hydrogenation processes, 35
  insertion mechanism, 76
  insertion reactions, 28–31
  interaction with zirconia, 124
  isotopes, 130
  methanation, 39
  petrochemical feedstock, 1
  pressure, 9,10*t*
  radical, 198
  reactions, 93*t*
  recovery
    economics, 6
    from natural gas, 3
  recovery technology, 5,6
  reduction
    catalysts, 81
    electrocatalytic, 52–90
    electrochemistry, 55–57
    formate oxidation, 176
    hydrogen insertion reaction, 147–154
    mechanisms, 53
    metal complexes, 53
    methanol, 55
    on metal electrodes, 178
    overpotential requirements, 55
    photosynthetic, 53
    products, 152*t*
    redox catalyst, 42–51
    thermodynamics, 54
  reduction pathways, 76
  solubility, 9,10*t*
  sources, 1–7,27
  supply, 5
  temperature, 9,10*t*
  traditional role, 1
Carbon dioxide–transition metal
  complexes, 52
Carbon monoxide
  competition, 31
  effect on methane formation rate, 160
  α-Carbon stereochemistry, 31
Carbon substrate, 159
Carbon surfaces, 156
Carbon-bound complexes, 59
Carbon–carbon coupling
  bound CO_2 molecules, 81
  oxalate, 57,61

Carbonate
  and bicarbonate species, 130
  desorption–decomposition, 125
  formation, 131
  ion, 60
Carbonic anhydrase
  active site, 97f
  amino acid residues, 96
  chemical properties provided by
    zinc, 99
  electronic reaction mechanism, 95–99
  metabolic role, 91
  second-order rate constants, 93t
  three-dimensionality, 100
Carbonium ion migration, 66
Carbonyl hydride, 135,139,141
Carbonylation vs. carboxylation
  reactions, 32t
Carboxylation of methanol, 35
Catalysis of CO_2, 92
Catalyst
  deactivation, 70,77
  high surface area carrier, 102
  iron oxide, 103,105t
  stability, 80
  systems, 81
  zinc oxide, 103
Catalytic activity
  design, 77–78
  iron oxide, 116
  metal on electrode surface, 81
  order, 112
  supported catalyst, 106
Catalytic principles, 99–101
Catalytic process
  mechanistic aspects, 37–39
  promoted by transition metal
    complexes, 33–39
Catalytic studies, 47
Catalytically active intermediates, 37
Cell compartment, 180
Chain length, 196
Chains of proteins, 101
Charge-separated state, 64
Chelate-type ligands, 83
Cluster-derived catalysts, 39
• Carbon monoxide dissociation, 168
Carbon monoxide hydrogenation
  oxygenated products, 139
  product distribution, 140t
  supported ruthenium catalysts, 134
Cobalt–phthalocyanine films, 68
Carbon dioxide hydrogenation
  effect of potassium, 141
  product distribution, 144t
  product species, 139,141
Coal-fired power plants, 5
Cobalt bipyridine complexes, 69
Contrathermodynamic step, 73

Coordination geometries
  CO_2 molecule, 16
  mixed carbon–oxygen coordination, 23
  pure carbon coordination, 19
  pure oxygen coordination, 21
Coordination of CO_2 to Ni, 16–25
Coordination states, 114–116
Cost of CO_2 recovery, 6
Coulombic stabilization, 23
Current–voltage curves, 159
Cyclic voltammetry
  CO_2 reduction, 156,158f
  combined with digital simulation, 78
  experiments, 47
  relative contributions of pathways, 76
Cyclic voltammogram, 71f,79f,82f

D

Dative bonds, 22f,24f
Deactivation pathway, 69,74
Decay kinetics, 62,63f
Deprotonated water molecules, 96
Desorption energy of CO_2, 120
Dielectric effect, 62
Digital simulation, 78
Dimer anion, 19
Direct insertion mechanisms, 83
Dissociative adsorption of CO, 164

E

Electrocatalysis, 147
Electrocatalysts
  CO_2 reduction, 52
  decrease overvoltage, 179
  stability, 76
Electrocatalytic CO production in
  CH_3CN solution, 78
Electrocatalytic reduction of CO_2, 66–80
Electrochemical cell, 183f
Electrochemical parameters, 45
Electrochemical potential, 153
Electrochemical reduction
  CO_2 to formate, 42–43
  CO_2 to formic acid and formate
    ions, 171
Electrochemical reversibility, 44
Electrochemical Stark effect, 200
Electrochemistry, 13
Electrode composition, 55
Electrode deactivation, 164,167
Electrode potential, 161,163t
Electrode ratios, 81
Electrodes
  carbon cloth, 72,77
  copper, 56

Electrodes–*Continued*
  modified, 80
  molybdenum, 56,172,174f
  palladium–hydrogen, 172
  platinum, 175,177f
  platinum gauze, 77,80
  precious metal, 176
  ruthenium, 56,173,174f
  ruthenium plate, 157f,166f
  ruthenium–iridium alloy, 177f
Electrolyte purity, 159
Electrolytes, 180
Electron density
  hydride, 28
  metal center, 29
Electron hopping, 116
Electron reservoir, 68,81,83
Electron transfer-catalyzed
    substitution, 77
Electron-protonation steps, 83
Electron-withdrawing CO_2 ligand, 80
Electronegative atom, 100
Electronegativity, 117
Electronic correlation, 16
Electronic rearrangements, 94
Electrophilic attack, 61
Electrophilic carbon, 28
Electropolymerization, 82f
Electroreduction of CO_2, 56
Electrostatic effects, 12
Enthalpy changes, 27
Entropy reduction, 95
Enzymatic activation, 91–101
Enzymatic catalysis, 93
Enzyme–substrate complex, 94
Enzymology, 95
Equilibria, 8
Equilibria of ion-pairing, 12
Equivalent circuit, 190
Etched surface, 182
Ethanol plants, 3
Ethylene, 56
Ethylene oxide, 4
Europium, 116,117
Evaluation of parameters, 190
Exchange current density, 196
Experimental methods
  adsorption on iron oxide, 103–106
  CO_2 interaction with zirconia, 124
  electrochemical reduction of
      CO_2, 156,180
  electrochemical reduction of
      CO_2 and HCOOH, 172
  electrostatic CO_2 reduction, 148–150
  rhodium catalysts, 134

F

Faradaic efficiency, 160,163
Film-based electrocatalyst, 81

Flue gas, 4,6
Formaldehyde, 55,150
Formate
  free, 75
  over ZnO, 123
  oxidation, 173,175t,176
  reduced to formaldehyde, 55
  reduction, 175t
Formate esters, 33
Formate ions, 173
Formate production, 72
Formate-producing pathway, 80
Formic acid, 57,150
Fused spheroids, 156

G

Generalized valence bond method, 16
Geometry of CO_2, 19

H

Half-wave potentials, 45
Hartree–Fock wave function, 21–23
Henry's law coefficient for CO_2, 10t
Hofmann degradation pathway, 72
Hole–electron pairs, 190
Homogeneous catalysis, 42
Hydride ligand, 28,76
Hydride transfer process, 76
Hydrogen
  absorption isotherm, 161
  evolution, 159
  gas
    effect on CH_4 formation rate, 161
    evolution, 72
    rate enhancement, 167
  insertion reaction, 147–154
  mass balance, 151
  plants, 4
Hydrogenation
  carbon monoxide, 123,135
  CO and CO_2, 133–146
  mechanism, 167–168
  reactions, 168–169
Hydrolysis, enzyme catalysis, 91
Hydrolytic stability, 65f
Hydroxyl groups, 125

I

Illumination, 13
Impedance
  bias potential, 185f,187f
  data, 198
  measurements, 181
  potential dependence, 185f

Impedance–*Continued*
  spectra
    CdTe–DMF, 182
    GaP–aqueous DMF, 186
    GaP–DMF, 187*f*
    tetraalkylammonium salts, 186
    total, 190,193
  Infrared spectra
    Fe/Al$_2$O$_3$, 109*f*
    Fe/SiO$_2$, 110*f*
    interaction of CO with Rh, 136*f*,137*f*
    interaction of H$_2$ and CO, 138*f*
    interaction of H$_2$ and CO$_2$, 142*f*,143*f*
    zirconium dioxide, 129*f*
  Infrared spectroscopy
    adsorption of CO and CO$_2$ on
      zirconia, 125
    adsorption on iron oxide, 104,108–116
    CO with supported Rh
      catalysts, 134–139
  Inhibition effect, 118,160
  Insertion reactions, 28
  Intermediate CO$_2$ complex, 74
  Internal electron transfer sites, 72
  Intramolecular acid–base interaction, 60
  Inverse isotope effect, 66
  Ion association, 12
  Iron oxide, 102–122
  Isobestic behavior, 62
  Isotope distributions, 128*t*

K

Kinetic barrier, 27,54
Kinetic studies, 47
Kinetics
  CO$_2$ insertion, 28
  water gas shift, 104

L

Labeling studies, 125,130
Light energy, 180
Lock and key noncovalent bonding, 92
Lone pairs, 17,23

M

Macrocycles, 67
Magnesium oxide, 102–122
Magnetite, 114
Many-electron wave
  function, 18*f*,20*f*,22*f*,24*f*
Mass signals, 126*f*,127*f*
Mechanistic selectivity, 67
Membrane potential difference, 151

Membrane system, 6
Metal alkoxides, 31–33
Metal alkyls and aryls, 28–31,43
Metal clusters, 68
Metal electrodes, 155
Metal hydride complexes, 49
Metal hydride foil, 148
Metal hydrides, 28
Metal oxides, 102–122
Metal porphyrins, 67
Metal–alkoxide bond, 63*f*
Metal–alkoxide insertion mechanism, 62
Metal–carbon bond distances, 29
Metal–hydride bond, 75,80
Metal–hydrogen bond, 43
Metal-induced transformations, 26–41
Metal–ligand bonds, 61–66
Metal–ligand charge transfer, 62
Metal–metal bonded dimeric species, 74
Metal–oxygen bond
  acid cleavage, 43
  strength, 117*t*,118
Metal–polypyridine complexes, 68
Metalloformate, 39
Methanation
  potassium poison effect, 141
  ruthenium clusters, 39
Methane
  formation activity, 160*t*
  formation mechanism, 155–170
  formation rate
    current vs. potential, 165*f*
    electrolysis time, 159,162*f*
    function of pH, 162*f*
    rate-limiting step, 168–169
    temperature, 165*f*
  steps in formation, 168*t*
Methanol
  formation rate, 151,152*f*
  reduction of CO$_2$, 55,56
Methyl formate production, 35
Miscible flood cooperation, 3
Mixed carbon–oxygen coordination, 23
Molecular mechanics computer programs, 101
Molecular orbitals, 17–23
Molecular sieves, 100
Molybdenum electrode, 172,174
Monodentate ligands, 83
Mössbauer spectroscopy, 103–116

N

Natural gas, 6
Nickel triad complexes, 45
Nitric oxide
  adsorbed, 111*t*
  adsorption probe, 103
  uptake, 105*t*,106
Nitrosyl species, 111–115
Nonaqueous media, 13

O

Oil recovery, 2
Ω-bonds, 17
One-carbon molecules, 26
Onset potential, 182
Oxalate, 56,61
Oxidation–reduction cycles, 120
Oxygen
  acceptor, 75
  hydrolysis, 99
  isotopes, 130
  labeling, 124
  sink, 72
  transfer, 60
Oxygenated products, 133

P

Palladium complex
  cyclic voltammetry, 47,48f
  NMR spectra, 46f
Palladium foil, 150
Palladium hydride bond, 49
Palladium membrane, 153
Palladium–hydrogen electrode, 172
Partial pressure of CO_2, 12
Passage of carriers, 190
Perfect-pairing orbitals, 17,18f
Petrochemical feedstock, 1
pH change during hydrolysis, 159
pH dependence of methanation, 167
pH effect on CH_4 formation rate, 161
Phase-transfer catalysis, 95
Phosphine ligand, 44,47
Photocatalyzed reduction of CO_2, 13
Photochemical reduction of CO_2, 66
Photocurrent–potential relationship, 182
Photoelectrochemical reduction
  p-CdTe, 180
  CO_2 on CdTe, 182,183f
  electron mediators, 201
  GaP, 186
  semiconductor electrodes, 57–58
Photosynthesis, 84
Phthalocyanines, 67
π-bonds, 17
Platinum
  absorbed CO_2^-, 191f
  adsorbed molecules, 202f
  CO_2 and CH_3CN, 191f
  counter electrode, 150
  electrode, 175,177f
  in acetonitrile
    CO_2^- radical adsorbed, 189f
    surface-adsorbed species, 188f
Polymeric electrocatalysts, 80
Polyoxymethylene glycols, 55

Polyphosphine metal complexes, 44–45
Polypyridine complexes, 52
Polypyridyl ligands, 68
Potassium
  CH_4 production, 145f
  effect on hydrogenation, 133–146
  electronic effect, 135
  poisoning effect, 141
Potential dependence, 182,186
Precious metal electrodes, 176
Product distribution, 133
Production of CO, 73
Proton source, 72
Protonation, 61,171

Q

Quaternary ammonium salts, 72

R

Rate limitation, 167
Rate of CO_2 insertion, 33
Rate-determining step, 179
Reaction with bases, 11
Redox catalyst, 42–51
Reduction mechanism, 75
Reduction of carbon dioxide
  Cobalt–macrocycles, 69
  products, 13
Refrigeration, 2
Reservoirs, 2
Resistance, 193
Resonance stabilization, 21
Resonance structures, 20f
Rhenium polypyridine complexes, 72
Rhodium catalysts, 133–146
Rhodium ions, 135
Ruthenium electrodes, 155–170
Ruthenium–iridium alloy electrode, 177f

S

Salting-out effect, 11
Scanning electron microscopy, 156
Second-order rate constants, 29t
Semiconductor electrodes, 57–58
Semiconductor electrolyte interface, 179
Sesquibipyridine ligand, 73
Side-bound coordination, 59
σ-bonds, 17
Silicon dioxide, 102–122
Simultaneous insertions of CO
  and CO_2, 31
Single-sweep techniques, 76
Sodium formate, 171–178

Solubility
  carbon dioxide, 9,10t
  nonaqueous solvents, 8
Solution pH, 55
Solvents
  dielectric constant, 13
  monoethanolamine, 11
  organic, 11
  propylene carbonate, 11
  water, 11
Space charge region, 193
Square-planar complexes, 45,68
Stabilization energy, 21–23
Stoichiometric reactions
  cobalt bipyridine complexes, 69
  redox catalyst design, 42–51
  transition metal complexes, 28
Structural types, 59
Support interaction, 102
Surface catalytic process, 167
Surface cations, 116
Surface spinels, 114
Surface-state density
  calculated from capacitance data, 197f
  GaP–electrolyte interface, 199f
Surface-state photoelectron transfer, 197f
Surface states
  capacitance, 194–199
  faradaic mediators, 196
  ionic adsorption, 194
  resistance, 196
Surface treatment, 182
Surfactants, 95

T

Temperature dependence
  adsorption on zirconium
    dioxide, 125–130
  effect of potassium, 139
  methane formation rate, 163t,164
  Rh–Al$_2$O$_3$ catalyst film, 135
Thermodynamically favorable
  reactions, 26–27
Titanium dioxide, 102–122
Titration alkalinity, 12–13
Transition metal complexes
  catalytic processes, 33–39
  reactivity, 58–67
Transition metal compounds, 27
Transition metal hydrides, 43
Transition state, 66

Triethanolamine solvent process, 6
Tungsten tetracarbonyl carbonate, 33
Tungsten–CH$_3$ bond distances, 31t
Turnover rate, 94
Two-bond maneuver, 99f
Two-electron pathway, 74
Two-electron reduction process, 70f,71f

U

Unsymmetric coordination, 23
Urea, 2

W

Wacker process, 42
Water gas shift
  activation energy, 120
  activity of bulk oxides, 117
  activity of Fe$_3$O$_4$/SiO$_2$ samples, 114
  iron and zinc cations, 106
  kinetics, 104,112,113t
  metal ion exchanged zeolites, 118
  reaction
    associative mechanism, 116,118
    effect of support, 103
    equation, 27
    pathway, 119f
    rate-controlling step, 118
    regenerative mechanism, 116
    reverse, 116–117
Water-based hydroxyl groups, 130
Wave function, 17,19

X

X-ray diffraction, 108

Z

Zeolites, 100
Zhang electronegativity scale, 117t
Zinc enzymes
  electronic rearrangements, 100
  ligands, 100
Zinc oxide, 102–122
Zinc-bound hydroxyl, 96
Zirconium dioxide, 123–132

Production and Indexing by Colleen P. Stamm
Jacket design by Carla L. Clemens

Elements typeset by Hot Type Ltd., Washington, DC
Printed and bound by Maple Press, York, PA

Recent Books

Personal Computers for Scientists: A Byte at a Time
By Glenn I. Ouchi
276 pp; clothbound; ISBN 0-8412-1000-4

The ACS Style Guide: A Manual for Authors and Editors
Edited by Janet S. Dodd
264 pp; clothbound; ISBN 0-8412-0917-0

Silent Spring Revisited
Edited by Gino J. Marco, Robert M. Hollingworth, and William Durham
214 pp; clothbound; ISBN 0-8412-0980-4

Chemical Demonstrations: A Sourcebook for Teachers
By Lee R. Summerlin and James L. Ealy, Jr.
192 pp; spiral bound; ISBN 0-8412-0923-5

Phosphorus Chemistry in Everyday Living, Second Edition
By Arthur D. F. Toy and Edward N. Walsh
362 pp; clothbound; ISBN 0-8412-1002-0

Pharmacokinetics: Processes and Mathematics
By Peter G. Welling
ACS Monograph 185; 290 pp; ISBN 0-8412-0967-7

Synthesis and Chemistry of Agrochemicals
Edited by Don R. Baker, Joseph G. Fenyes, William K. Moberg,
and Barrington Cross
ACS Symposium Series 355; 474 pp; 0-8412-1434-4

Nutritional Bioavailability of Manganese
Edited by Constance Kies
ACS Symposium Series 354; 155 pp; 0-8412-1433-6

Supercomputer Research in Chemistry and Chemical Engineering
Edited by Klavs F. Jensen and Donald G. Truhlar
ACS Symposium Series 353; 436 pp; 0-8412-1430-1

Sources and Fates of Aquatic Pollutants
Edited by Ronald A. Hites and S. J. Eisenreich
Advances in Chemistry Series 216; 558 pp; ISBN 0-8412-0983-9

Nucleophilicity
Edited by J. Milton Harris and Samuel P. McManus
Advances in Chemistry Series 215; 494 pp; ISBN 0-8412-0952-9

For further information and a free catalog of ACS books, contact:
American Chemical Society
Distribution Office, Department 225
1155 16th Street, NW, Washington, DC 20036
Telephone 800-227-5558